AF541076

Plant Biodiversity and Taxonomy

Plant Biodiversity and Taxonomy

by

Dr. M.P. Singh, *FMA*

Dr. B.S. Singh

Soma Dey

2024

Daya Publishing House®

A Division of

Astral International Pvt. Ltd.

New Delhi – 110 002

First Published, 2002

ISBN: 978-81-7035-289-1 (HB)

Published by : **Daya Publishing House®**
A Division of
Astral International Pvt. Ltd.
– ISO 9001:2015 Certified Company –
4736/23, Ansari Road, Darya Ganj
New Delhi-110 002
Ph. 011-43549197, 23278134
E-mail: info@astralint.com
Website: www.astralint.com

Digitally Printed at : Replika Press Pvt. Ltd.

FOREWORD

Biological diversity has a long history of usage in a variety of context, but the start of its rise in the current senses can be raced to three publications which appeared in 1980 : Love Joy (1980 a,b) did not provide a formal definition but used it essentially in the sense of the number of species present, and Norse & McManus (1980), employed it to include two related concepts : "Genetic diversity" and "ecological diversity" The latter, it has been equated ecological diversity with species richness, the number of species in a community of organisms. For practical purposes biodiversity can be considered as Synonymous with 'biological diversity' as defined by (Norse *et. al.* 1986). This is reinforced by official definition in Article 2 of the convention of Biological diversity, signed by 156 nations and the European community at the United Nations conference of the Environment and Development on the Earth Summit in 1992 which closely mirrors the concept of Norse *et. al.*

A number of higher taxa, for example Phyla, orders and families provide a more appropriate measure of the biodiversity as in a site than the number of species.

The scope of this publication is limited to the community and species diversity and is intended to provide a full and over all view of the taxonomical diversity in the country. This book has very wide and should provide help to all Conservationist, Environmentlists and Taxonomists of the country. The authors are praise worthy for this publication.

Prof. D.L. Hawksworth
Director,
International Mycological Institute,
Surrey, United Kingdom

PREFACE

India has an age old tradition of nature conservation and the same is reflected not only in old literature and cultural ethics but also in the constitution, policies, legislation and organisations.

Biodiversity forms the root of all living system. India is *fortunate* enough to be ranked sixth among the twelve *mega* biodiversity countries. Its biological resources include 50,000 species of plants and 81,000 species of animals including ones belonging to the lower Phylas. However due to habitat loss and over exploitation owing to burgeoning population. The biodiversity of our country is severely threatened and some species which one abundantly found, have now become rare and some have become extinct. The Indian Cheetah (*Acinonyx Jabatus*) is one such stark example. The concern of the Government towards the conservation of biodiversity has grown considerably. This is amply reflected in the Wildlife (Protection) Act, 1972, Forest Conservation Act, 1980 and the Forest Policy of 1988. The Indian Government is also a party to a number of International treaties and conventions which include the CITES, Ramar convention, Biodiversity convention, etc. of late, the modalities of Patenting of biodiversity under intellectual Property Rights (IPR) are being worked out to check the piracy of genetic material. A number of institutions in the world are working towards both *exsitu* and *insitu* conservation of biodiversity. The conservation of biodiversity is linked with the maintenance of ecological Stability and Productivity. These *attributes* are important in sustained development and stable national economy (McNeely, 1988). The concept of biodiversity is very vast and seated at a microlevel. In holistic sense, it can be linked with most of the biological processes in nature. Its conservation involves a number of parameters such as number of species, their Population dynamics, distribution, habitat, structure, micro habitats, Physical environments climate,

Present management and past history. Therefore, the efforts for conservation of biodiversity should be in tune with the processes and its occurrence in space and time, from microlevel to megalevel. The problem of biodiversity conservation has become a global issue. It is being realised that the forest existing in a country is not a resource just for that country, but for the whole of the world. The Amazonian Rain Forests have been called the "Lungs of the World" as they serve to puity the global atmosphere by release of oxygen and obsorption of pollutants. The present book is based on numerous materials, reports personal communication with scientists naturalists and author's own extensive surveys and researches on the flora of various parts of the nations. The authors are also fully conscious of the several limitations in this work, particularly lack of sufficient field work and study of some Phytogeographically to significant groups. It is hoped that the publication will be welcomed by all taxonomists, foresters, environmentalists, naturalists, Politicians and other decision makers of the nations.

A number of people have helped us in different means for going through the publication. We are particularly thankful to Professor H.Y. Mohan Ram, Retd. Professor of Botany, Delhi University, Delhi for encouraging us to take up this work, Dr. A.K. Sharma, Professor Eminence, Indian National Science Academy, New Delhi for correction of manuscripts of this work. We are thankful to a number of teachers, Scientists for their assistance. Dr. R.N. Trivedi, Dr. K.K. Nag, Dr. N. Dayal, Dr. S.N. Dixit, Dr. S.N. Sachan, Dr. P.K. Khosla, Dr. D.N. Tiwary have kindly spared valuable discussion on some issues. We would be failing in our duties if we do not record our special thanks to our children, who have been a constant source of encouragement during the entire course of this work.

Ranchi

M.P. Singh
B.S. Singh
Soma Dey

CONTENTS

Introduction

In India, out of the expected 15,000 to 16,000 species of flowering plants about 1500 (10 per cent) have already come under the various categories of threatened plants (Appendix 2). According to Nayar (1989) out of 427 endangered species published in the Red Data Books of India (Nayar and Sastry, 1987-90) it is noticed that 28 species are supposed to be extinct, 24 as endangered, 81 as vulnerable and 160 as rare and 34 as insufficiently known. The monotypic endemic genus Hubbardia (Poaceae) described from Job falls of Karnataka is now said to be extinct. In Andaman and Nicobar Islands, about 65 endemic species, mostly tree species could not be relocated during recent botanical explorations by the Botanical Survey of India. Similarly, during the last fifty years about, 100 species could not be recollected from their type localities. Unless the habitat loss and population pressure is curtailed we may have to say Good Bye to another 100 species by the turr of the century itself.

The alarming rate of loss of biodiversity in India is certainly due to the percentible decline in the forest cover and loss of species specific habitats over the years.

The causes that have posed a serious threat to the forest cover in India are both natural and man-made. The natural causes responsible for the threat to flora include floods, earthquakes,

Landslides, natural competition between the species, biology of species such as lack of pollinators and natural regeneration, diseases and spread of alien weedy species. The man-made threats include mainly expansion of agriculture, harvesting of fuelwood and timber, grazing in forest lands, forest fires, exploitation of forest products and selective removal of species and finally the developmental activities, all at the cost of the natural flora.

The increase in human population during the last few decades is demanding development in all spheres including agriculture. There is a vast expansion of agricultural land to meet the growing need of mankind. Clearing forests for Agriculture is a historical phenomenon in India. The beginning of agriculture was in the form of shifting cultivation (known as Jhum in most parts of India). But this age-old practice is rooted even today in the cultural and religious ethos of several tribal communities shifting cultivation is one of the most important factors in the degradation of the country's floristic diversity.

Fuelwood is an important source of energy for Indian rural households, which make up *ca.* 75 per cent of population. In fact, 80 per cent of India's wood production goes to fuelwood, 18 per cent to timber and only 2 per cent to industry. The present firewood consumption of fuelwood is estimated at 135 million tonnes in rural India and 23 million tonnes in Urban area, much of this coming from the natural forests.

Grazing in forest land has also led to the excessive removal of undergrowth in the forests. It is estimated that currently over 90 million animals graze in the forests and as much as 83 per cent of forest land is subject to grazing. The forest fires, in no way cause less damage to the forests. Once a forest fire breaks, it spreads for miles together killing the entire biodiversity of the area.

Exploitation and selective removal of certain group of plants from the forests have endangered these groups. For example *Dioscorea deltoidea* Wall. (diosgenin bearing) which occurs in Kashmir, Himachal Pradesh and hill region of Uttar Pradesh has been indiscriminately exploited for at least 20 years and in some areas it is now virtually extinct. The same is true for *Podophyllum hexandrum* Royle, *Panax pseudo-ginseng* Wall., *Coptis teeta* Wall. and *Rauvolfia serpentina* Benth. ex Kurz. Similarly minor forest products like some commercially valuable gums (*Sterculia urens* Roxb.), cutch (*Acacia catechu* Willd.), tannins (*Terminalia* and *Emblica* spp.), for brooms (*Thysanolaena maxima* (Roxb.) Kuntze), tendu leaves (*Diospyros* spp.), essential oils (*Cymbopogon, Vetiveria* spp.). The roots of *Vetiveria zizanioides* (L.) Nash are rich in aromatic oil and the rural folk exploit this in bulk quantity for making screens and curtains, which are extensively sold in Uttar Pradesh during summers.

In addition to the above, developmental activities such as the growth of large Urban areas, construction of dams, buildings, road building on hills and forested areas, mining operations, etc., are all examples of direct onslaughts of nature resulting in the loss of biodiversity.

But the situation is not too hopeless. Despite the shrinkage in population of many species and extinction of some of them, the vast majority are still available in the country. It is therefore, necessary to conserve the biodiversity available in the country. There has been a greater awareness towards the environment in the country since the early 70's. The fourteen Biosphere Reserves in the different climatic zones and the numerous other protected areas like the wildlife sanctuaries, National Parks, etc., covering approximately 3.5 per cent of the country can certainly take care of the *in situ* preservation of the rich floristic diversity. Efforts can also be made in the coming years to increase the protected area network to cover atleast 5 per cent of the country. In addition to these legally protected areas, every city, town and village must also develop their own chunk of forest as 'heritage sites', 'wilderness areas', 'sacred forests' or 'sacred woods' or 'community gene banks' on similar lines of Mawphlang, a small town in Meghalaya. Such wilderness areas can be developed without much inputs in terms of money and man-power. Even school children can take a lead role in such programmes. While such vegetation zones in the viscinity of urban areas greatly improve the quality of environment polluted by the concrete jungles, automobile exhaustion and sewage disposal, and at the same time enrich the plant diversity of the country.

Chapter 1
Plant Biodiversity

It is now well realised that due to some natural and unnatural factors, the biological resources of this earth are under varying degrees of threat, and more so in the tropical regions. In India, during the last two decades there has been a greater consciousness about the threat to the biological species and it is widely realised that the loss of biological diversity has a great economic and environmental impact. It is roughly estimated that *ca.* 10 per cent of the flowering plant species in India are threatened. While conservation and multiplication of these species is to be attempted, all efforts are to be diverted towards the proper conservation and wise management of the remaining 90 per cent of the flowering plant species that we still have at our disposal. For launching any such policies and programmes on conservation and sustainable development of the resources, it is necessary that proper assessment is made of the floristic diversity prevailing in the country. Correct inventorisation and assessment of biodiversity *in different habitats* is also necessary for evolving a long term strategy for rehabilitation of the endangered species in similar, alternate habitats when original habitat gets destroyed.

Diversity simply means to be different or unlike. Floristic diversity refers to the groups or classes of plants to be varied. In other words, it refers to the number of types or taxa in a given region or a group. The concept of biological diversity is defined as the variety of life forms in a given region, the ecological roles they perform and the genetic diversity they contain (Wilcox, 1984). Biological diversity can be measured at any level from overall global diversity to ecosystem, community, species, population, individuals and even to genes within a single 'individual'. A region with a large number of individual species is said to be biologically diverse, where the number of groups of taxa and individual species provide the criteria for measurement of biodiversity. On the other hand, a species with a large number of inter populational differences, where each population gets genetically adapted to specific environmental conditions is said to be genetically diverse, i.e. genetic diversity of a species. In India, studies on the diversity at the species and infraspecific level are too meagre, though such studies are urgently required for the biological conservation of a species.

The main theme of the book is to highlight the floristic diversity in the former perspective, i.e. diversity in terms of species richness. In doing so the author is fully conscious of the fact that the number of individual species alone may not determine the diversity of a region or group. For example, the genus *Syzygium* of Myrtaceae has maximum concentration of species (over 200 species) in Malaysia, all of which are more or less closely related and thus; show less genetic diversity; whereas the 52 Australian species show greater morphological differences and hence greater genetic diversity (Hyland, 1983). Nevertheless, species number in a community/ region certainly plays a very important role in prioritising the diversity centres for purposes of conservation.

The Indian Region

The Indian region is one of the most distinct biogeographic regions of the world. India is the seventh largest country in the world and Asia's second largest country with a total area of 3029 million hectares. Being one of the most populous regions of the world India is primarily an agriculture dominated country. The Indian mainland lies between 6°45′ to 37°6′N latitude and 68° 7′ to 97° 25′ E longitude and has a land frontier of *ca.* 15,200 km and a coastline of 5400 km. In the north it is bounded by Tibet, Nepal and Bhutan; in the north-west by Pakistan; in the north-east by Myanmar (formerly Burma); and in the east by Bangladesh (formerly East Pakistan). The southern peninsula extends into the

Indian Ocean with the Bay of Bengal lying to the south-east and the Arabian sea to the south-west.

Physiographically, the Indian mainland may be divided into three distinct regions–the Himalaya, the Indo-Gangetic plain and the Peninsular India or the Deccan region. The Himalaya occupying the extreme northern margins of India has a great influence on the climate of the entire region. The Himalayas comprise a complex series of geologically recent folded mountain ridges interspersed with extensive plateaux and include some of the highest peaks of the world. The highest peak in the Indian Himalaya lies in Kanchenjunga (8586 m) which is located in Sikkim on the border with Nepal. The Himalayan mountain ranges extend east-west from the River Indus to the River Brahmaputra over 2400-2500 km long and with an estimated width varying between 160 - 400 km. They are a continuous chain of three longitudinal mountain zones, namely

(a) The Great Himalaya or the Trans Himalaya in the north with the average elevation of 6000 m.

(b) The Lesser Himalaya or the middle Himalaya with the average elevation of 4500 m.

(c) The Outer Himalaya or the Siwalik ranges from the Gangetic Plains to 1200 m elevation.

Paralleling the mountain wall to the south are the low lands of the alluvial indo-Gangetic plain. The Indo-Gangetic plain stretches from Assam in the east to the Punjab in the west (*ca.* 2400 km). Some of the largest rivers of India including Ganga, Yamuna and Brahmaputra flow across this region. Topographically the Indo-Gangetic plain is remarkably homogenous except for the low hill ranges, the Siwaliks and the Aravallis extending north-east to south-west. A greater part of the region with the alluvial silts and clays of the major rivers is agriculturally highly productive and strongly contrast with the almost arid and sterile sands of the Indian desert at the extreme west of the Rajasthan state.

The Peninsular India (Deccan Plateau) in the south and the warm alluvial Indo-Gangetic plains in the north are separated roughly by the boundary of the Tropica of Cancer. Peninsular India (22° N to 8° N) is dominated by the great triangle of the eastward tilting Deccan plateau, and Archaean shield some 39,000 km^2 in extent. The Peninsular India consists of ancient crystalline rock formation which have been worn down to form a series of plateaux. These generally slope eastwards and are drained by large rivers such as Mahanadi, Krishna, Godavari, Tapati and a few others. The Peninsula is flanked on one side by the Eastern Ghats (average elevation 600 m) and on the other side by the Western Ghats (1000-2400 m elevation). Between the Western Ghats and the Arabian sea and between the Eastern Ghats and the Bay of Bengal lie a narrow coastal plain with a unique Climate and biota. The Eastern and Western Ghats meet in the Nilgiri hill region of Karnataka and Tamil Nadu on the southern tip of the Indian Peninsula.

According to the continental drift theory, the Peninsular India represents the Indian part of the Gondawanaland, which on breaking some 1000 m. Y.B.P. during Cretaceous era moved northwards and crashed against Laurasia causing the upliftment of the Himalaya and obliteration of the former Tethys Sea. In the process of movement of Indian plate over 5,900 km passing through different climatic regions followed by volcanic eruptions, causing the flow of deccan lavas, major groups of the original Gondawanaland floristic stock became extinct. The upliftment of the Himalaya created a vast chain of events resulting in the shaping of land formations and river systems, which finally resulted in the formation of the alluvial Indo-Gangetic plain. These geological events resulted in the evolution of flora, extinction of vulnerable groups due to climatic changes, migration of flora through new corridors of mountain chains, adaptive radiation of species complexes in the new ecological niches and refugiums afforded by these mountain systems.

Climate

The climate of India is extremely varied. There are some of the highest mountain peaks covered by perpetual now and in contrast to this there are also some of the hottest placed on earth. There are almost zero rainfall areas in the extensive dry desert of the west to the world's highest rainfall areas in the north-east India. The mountain ranges in the north exert a major influence on the climate of India. The northernmost parts of India reveal a colder climate, often classified as montane temperate. A larger part of the Deccan Peninsula and north-eastern part of the country present a tropical climate. Central and northern India are nearly subtropical with a strong seasonality of the climate. In the Himalayan foothills the climate is warm temperate or montane temperate. No doubt, this extreme variation in climate has resulted in an enormous floristic diversity in the country.

The climate of India is dominated by the Asiatic monsoon, a great wind system that seasonally reverses or changes their direction resulting in three well marked seasons, particularly in the northern India. From June to October the rains come from south-west while between December and February the monsoon blows from north-east. Seasonality of rainfall is thus a conspicuous feature of the Indian climate. On touching the southern end of the Peninsula, the monsoon winds over the Indian Ocean bifurcate into two directions, one of which moves from the south-west over the west coast of India towards north-east and the other passes over the Bay of Bengal towards Bangladesh and north-east region. As the monsoon winds from the Bay of Bengal get trapped between the high hills of the north-east, it causes a very heavy downpour and then it is deflected westwards, where, by the time the monsoon winds reach the north-west part, the moisture gets completely depleted, thereby causing very litle rain. By early October the south-west monsoon starts to retreat. Under the influence of monsoons the rainfall decreases in general from east to west. Maximum rainfall occurs over parts of north-east India (Cherrapunji in Meghalaya with an annual rainfall of 1200 to 1300 cm is the world's wettest place) and along the Western Ghats. Much of the western part of the Deccan plateau, east of the Western Ghats lies in the rain shadow zone and hence receives relatively low rainfall.

Climatically, the country may be divided into six zones–the northern mountains, northern plains, Deccan plateau, west coast region, south-east coastland and extreme north-east. In northern mountains rain may fall at any time although the main rainy season is during July to September. In the northern plains rainfall decreases from east to west with the true desert conditions prevailing in the extreme west of Rajasthan. Rainfall in the interior region of Deccan plateau is determined by the seasonal changes of the south-west monsoon. The west coast region is bounded to the west by the Arabian sea and to the east by the Western Ghats. Here, the rains occur more or less throughout the year with the temperature and humidity remaining high during the monsoon. The rainfall in the south-east coastland is generally associated with cyclones and storms which develop in the Bay of Bengal during October to December. Rainfall is lighter, although temperature and humidity remain high. The climate of extreme north-east is similar to that of the northern plains except for the heavy rainfall.

Temperature variation in different regions and during different seasons is also very great. The summer temperature in most of the regions rises to above 38° C. The minimum winter season temperature in southern India rarely go below 20°C and as one moves northwards the winter mean temperature falls lower and lower with about 4°C in the foothills and in the Kashmir valley. The lowest recorded temperature in the north-western parts of India is about 8°C. Rarely winter frost is also observed in parts of Punjab, Himachal Pradesh and northwards (Gopal, 1990).

Soils

Soil is one of the most important ecological factors which determines the distribution of the ecological groups of plants, such as Oxylophytes (on acidic soils) Halophytes (on saline soils), Psammophytes (on sand), Lithophytes (on rock surfaces) and Chasmophytes (in rock crevices) as proposed by Warming (1909). Mainly four types of soils – alluvial, black, red and lateritic soils are predominant in the region (Raychaudhuri *et. al.*, 1963).

Alluvial soils are derived from the sand and silt deposited by the rivers of the Indo-Gangetic plains and occupy a major area of the country. They often countain large amounts of calcium in the form of calcium carbonate granules resulting in waterlogging. Most alluvial soils also have a high clay content. Black cotton soils or regur are derived chiefly from the Deccan trap in Peninsular India. They are generally fertile and have a fine texture. Red soils are derived from ancient metamorphic rocks and are poor in nitrogen, phosphorus and organic contents. Red soils mostly occur in the eastern part of Peninsular India. Lateritic soils which occur mostly on the hills of Peninsular India are rich in hydrated oxydes of aluminium and iron. There are many other minor types of soils which have a restricted distribution in India. Saline and alkali soils occur in many parts of the alluvial plains. Desert soils are noticed in the arid and semi-arid parts of the country. Forest soils with a very high organic matter content exist in areas covered by deciduous forests.

Floristic Diversity

The richness of the vegetational covering of India is well known. The Indian region (6° 45' to 37° 6' N and 68° 7' to 97° 25' E) with a total area of about 3029 million hectares is considered to be one of the twelve centres of origin and diversity of several plant species in the world. According to the Russian botanist N.I. Vavilov (1926) the Indian region forms the 'Hindustan centre of origin of cultivated plants'. The great British botanist, Sir J.D. Hooker (1904) remarks at one place "The Indian Flora is more varied than that of any other country of equal area in the eastern hemisphere, if not on the globe". It is estimated that about 17,000 vascular plants (flowering plants, gymnosperms and pteriodophytes) are accounted for in this region.

A significant feature of the Indian flora is the confluence of floras from the surrounding countries like Malaya, Tibet, China, Japan, Europe and even from wide separated continents like America, Africa and Australia. This fact prompted Hooker (1904) to arrive at the erroneous conclusion that India has no flora as a separate entity but is an admixture of the floras from the adjacent countries. The subsequent phytogeographers after critical analysis of flora have convincingly concluded that India has a flora of its own, and infact as many as 5,000 species are endemic to this region (Chatterjee, 1940, 1962; Nayar, 1980).

India's rich vegetational wealth and diversity is undoubtedly due to the immense variety of the climatic and altitudinal variations coupled with varied ecological habitats. There are almost rainless areas to the highest rainfall area in the world. The altitude varies from the sea level to the highest mountain ranges of the world. The habitat types vary from the humid tropical Western Ghats to the hot desert of Rajasthan; from cold desert of Ladakh and icey mountains of the Himalayas to the long, warm coast line stretches of Peninsular India. The extreme diversity of the habitats has resulted in such luxuriane and variety of flora and fauna that almost all types of forests, ranging from scrub forest to the tropical evergreen rain forest, coastal mangrove to the temperate and alpine flora occur in this region. Champion and Seth (1968) have recognised 16 major forest types comprising 221 minor types in India (Table 1.1). Of these, the Tropical moist deciduous forest forms the major percentage (37%) of forest cover in India. Tropical dry deciduous forest forms 28.6% and the remaining ones are scattered in minor percentage. The officially recorded forest area in the country is 75.18 million ha, although the Satellite imagery puts the figures at 64.2 million ha. The richness and diversity of flora of India can be further appreciated by the fact that as many as 10 biogeographic regions representing 3 basic biomes and 2 natural realms as identified by Udvardy (1975) are recognised within the territory of the Indian Republic. These are :

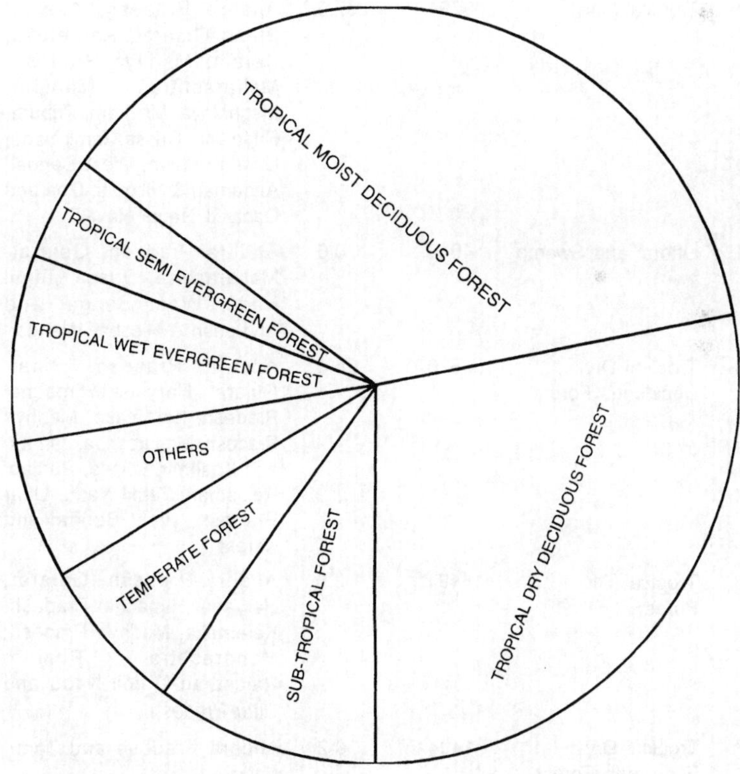

Fig. 1.1 : Forest Type Distribution

Source : *ANONYMOUS* 1987b

Table 1.1 Major Forest Types in India (After Champion & Seth, 1968)

Sl. No.	Forest Type	Area in sq. km.	Percentage	Statewise Occurrence
1.	Tropical Wet Evergreen Forest	51249	8.0	Arunachal Pradesh, Assam, Karnataka, Kerala, Manipur, Nagaland, Tamil Nadu, Andaman & Nicobar Islands and Goa.
2.	Tropical Semi-Evergreen Forest	26424	4.1	Assam, Gujarat, Karnataka, Kerala, Maharashtra, Nagaland, Orissa, Tamil Nadu, Andaman & Nicobar Islands and Goa.
3.	Tropical Moist	236794	37.0	Andhra Pradesh, Assam, Bihar, Gujarat, Karnataka, Kerala, Madhya Pradesh, Maharashtra, Manipur, Meghalaya, Mizoram, Tripura, Nagaland, Orissa, Tamil Nadu, Uttar Pradesh, West Bengal, Andaman & Nicobar, Goa and Dadar & Nagar Haveli.
4.	Littoral and Swamp	4046	0.6	Andhra Pradesh, Gujarat, Maharashtra, Orissa, Tamil Nadu, West Bengal and Andaman & Nicobar Islands.
5.	Tropical Dry Deciduous Forest	186620	28.6	Andhra Pradesh, Bihar, Gujarat, Haryana, Himachal Pradesh, Karnataka, Madhya Pradesh, Maharashtra, Jammu and Kashmir, Orissa, Punjab, Rajasthan, Tamil Nadu, Uttar Pradesh, West Bengal and Kerala.
6.	Tropical Thorn Forest	16491	2.6	Andhra Pradesh, Gujarat, Haryana, Himachal Pradesh, Karnataka, Madhya Pradesh, Maharashtra, Punjab, Rajasthan, Tamil Nadu and Uttar Pradesh.
7.	Tropical Dry Evergreen Forest	1404	0.2	Andhra Pradesh and Tamil Nadu
8.	Sub-Tropical Broad Leaved Hill Forest	2781	0.4	Assam, Maharashtra, Meghalaya and West Bengal

Contd...

Table 1.1 contd...

Sl. No.	Forest Type	Area in sq. km.	Percentage	Statewise Occurrence
9.	Sub-Tropical Pine Forest	42377	6.6	Arunachal Pradesh, Himachal Pradesh, Jammu & Kashmir, Manipur, Meghalaya, Nagaland, Sikkim and Uttar Pradesh
10.	Sub-Tropical Dry Evergreen Forest	12538	2.5	Himachal Pradesh, Jammu & Kashmir and Mizoram.
11.	Motane Wet Temperate Forest	23365	3.6	Arunachal Pradesh, Karnataka, Manipur, Nagaland, Sikkim and Tamil Nadu.
12.	Himalayan Moist Temperate Forest	22012	3.4	HImachal Pradesh, Jammu & Kashmir and Uttar Pradesh.
13.	Himalayan Dry Temperate Forest	312		Jammu & Kashmir and Himachal Pradesh.
14.	Sub-Alpine	18628	2.9	Arunachal Pradesh, Himachal Pradesh, Jammu & Kashmir. Nagaland, Sikkim and Himachal Pradesh
15.	Moist Alpine			Jammu & Kashmir, Uttar Pradesh and Sikkim
16.	Dry Alpine Scrub			Jammu & Kashmir

1. Himalayan highlands.
2. Thar desert
3. Malabar rain forest
4. Indo-Ganges monsoon forest
5. Deccan thorn forest
6. Coromandel
7. Mahanadian
8. Bengalian rain forest
9. Laccadive Islands
10. Andaman and Nicobar Islands

It is estimated that over 45,000 species of plants are accounted for in this region which represent 11 per cent of the known plant species of the world. These are distributed in the following groups :

Table 1.2

Division	Approximate No. of species
Angiosperms	15000
Gymnosperms	64
Pteridophytes	1022
Bryophytes	2584
Lichens	1600
Fungi	23000
Algae	2500
Bacteria	850

The flowering plants of India comprise about 15000 species and represent 6 per cent of the world's known flowering plants (Nayar, 1977). About 315 families (of *ca.* 400 now defined) and 2250 genera of flowering plants are known to occur in India in different ecosystems from the humid tropics of Western Ghats to the alpine zones of the Himalaya and from mangroves of tidal Sunderbans to the dry deserts of Rajasthan. The ten largest families by number of species in order of sequence are as shown in Table 1.3.

Table 1.3

	Name of the family	*No. of genera*	*No. of species*
1.	Poaceae (Gramineae)	255	1255
2.	Orchidaceae	145	990
3.	Fabaceae (Papilionaceae)	123	775
4.	Asteraceae (Compositae)	161	1000
5.	Rubiaceae	90	495
6.	Cyperaceae	24	449
7.	Euphorbiaceae	74	419
8.	Lamiaceae (Labiatae)	68	393
9.	Acanthaceae	84	379
10.	Scrophulariaceae	66	356

In addition to this, several families also show great floristic diversity and are represented by more than 100 species. These are listed in Table 1.4.

Table 1.4

Name of family	*No. of taxa*	*Name of family*	*No. of taxa*
Ranunculaceae	160	Lauraceae	163
Brassicaceae	164	Euphorbiaceae	419
Caryophyllaceae	110	Moraceae	105
Malvaceae	100	Urticaceae	115
Fabaceae	775	Ericaceae	150
Rosaceae	217	Lamiaceae	393
Apilaceae	202	Scrophulariaceae	356
Rubiaceae	495	Acanthaceae	379
Asteraceae	1000	Orchidaceae	990
Primulaceae	165	Zingiberaceae	133
Asclepiadaceae	209	Liliaceae	159
Gentianaceae	147	Arecaceae	90
Boraginaceae	138	Araceae	138
Convolvulaceae	161	Cyperaceae	449
Myrsinaceae	115	Poaceae	1225

Several taxa such as *Berberis* have in this region a primary centre of diversification. These taxa reveal maximum diversity not only in different ecological zones but also in one and the same habitat, often posing immense taxonomical problems. The genus *Berberis* can be cited as a good example under this category. This genus of Himalayan region has *ca.* 52 species with *ca.* 30 infraspecific categories revealing a high degree of diversity.

Table 1.5 Some taxa of Berberidaceae having maximum diversity in the HImalaya

Species	Recognised infraspecific categories
Berberis insignis	*insignis, elegantifolia, zelaica, tongloenis, shergaonsis, gouldii*
B. lycium	*lycium, subfascicularis, simlensis, subvirescens*
B. orthobotrys	*orthobotrys, canescens, sinthanensis, conwayi*
B. hookeri	*hookeri, viridis, platyphylla, microcarpa*
B. macrosepala	*macrosepala, Sakdenensis, setifolia*
B. umbellata	*umbellata, brainii*
B. sublevis	*sublevis, microcarpa*
B. asiatica	*asiatica, clarkeane*
B. angulosa	*angulosa, fasciculata*
B. concinna	*concinna, brevior*
B. chitria	*chitria, occidentalis*
B. coriaria	*coriaria, patula*
B. jaeschkeana	*jaeschkeana, usteriana*
B. pachyacantha	*pachyacantha, zabeliana*
B. pseudumbellata	*pseudumbellata, gilgitica*
B. petiolaris	*petiolaris, garhwalana*
B. koehneana	*koehneana, auramea*

On the other end of the spectrum, there are several monotypic families. Over 60 families are believed to be represented so far by just one species in India, e.g. Coriariaceae, Turneraceae, Illiciaceae, Ruppiaceae, Philydraceae, Tetracentraceae and Siphonodontaceae, etc.

Of the 15000 species of flowering plants, about 2560 (17%) species belong to the tree species and they predominantly occur in the following families.

Table 1.6

Name of family	No. of tree species
Euphorbiaceae	200
Lauraceae	150
Annonaceae	100
Rubiaceae	100
Moraceae	100
Fabaceae	100
Rubiaceae	80
Arecaceae	80
Meliaceae	71
Mimosaceae	60
Caesalpiniaceae	60

Some of the highly valued timber species of the world hail from this region. The distribution and quality of these timbers vary from region to region. A few highly reputed timbers of India are listed in Table 1.7.

Table 1.7 Important timber species of India

Species	Trade Name
Albizia odoratissima	Black siris
Terminalia bellirica	Bahera
Lagerstroemia mircocarpa	Benteak
Phoebe sp.	Bonsum
Bischofia javanica	Bishop wood
Tectona grandis	Teak
Michelia champaca	Champ
Schima wallichii	Chillnni
Cinnamomum spp.	Cinnamon
Artocarpus chama	Chaplash
Chukrassia velutina	Chickrassy
Dillenia spp.	Dillenia
Dipterocarpus kerrii	Gurjan
Gmelina arborea	Gamari
Abies pindrow	Silver fir
Ailanthus integrifolia	Gokul
Terminalia myriocarpa	Hollock
Haldina cordifolia	Haldu
Dipterocarpus macrocarpus	Hollong
Hopea glabra	Hopea

Contd...

Table 1.7 – *contd...*

Species	Trade Name
Terminalia chebula	Harra
Syzygium spp.	Jamun
Altingia excelsa	Jutili
Castanopsis purpurilla	Indian chestnut
Mesua floribunda	Karol
Mesua ferrea	*Mesua*
Anthocephalus chinensis	Kadam
Careya arborea	Kumbi
Carallia brachiata	Manigewa
Mangifera indica	Mango
Shorea assamica	Makai
Persea macrantha	Machilus
Azadirachta indica	Neem
Maniltoa polyandra	Ping
Aphanamixis polystachya	Pitraj
Stereospermum dereonatum	Padri wood
Lagerstroemia hypoleuca	Pyinura
Shorea robusta	Sal
Dalbergia sissoo	Sheesham
Toona ciliata	Toon
Dysoxylum malabaricum	White ceder
Albizia procera	White siris
Juglans regia	Walnut
Cedrus deodara	Deodar, Himalayan Ceder
Pinus roxburghii	Chir Pine
Acacia catechu	Cutch tree
Bombax ceiba	Silk cotton tree
Holoptelea integrifolia	Papri, Chilbil
Boswellia serrata	Indian olibanam tree
Ougeinia oojeinensis	Sandan
Quercus dilatata	Green oak
Q. leucotrichophora	Ban oak, Grey oak
Q. semicarpifolia	Brown oak
Schleichera oleosa	Lac tree
Stereospermum suaveolens	Padal
Zanthoxylum alatum	Tumru wood
Lannea coromandelica	Jhingan
Olea ferruginea	Indian olive
O. glandulifera	–

Contd...

Table 1.7 – *contd...*

Species	Trade Name
Rhamnus virgata	Indian Buckthorn
Salix alba	White willow
Wrightia tinctoria	Pala Indigo-plant
Grewia optiva	Bhimal tree
Garuga pinnata	Kharpat wood
Fraxinus hookeri	Hooker ash
Ehretia acuminata	Puna
Anogeissus latifolia	Dhawa
Alnus nepalensis	Indian Alder
Alstonia scholaris	Dita bark wood
Picea smithiana	Himalayan spruce
Aesculus indica	Indian horse-chestnut
Pistacia integerrima	Kakra wood
Populus ciliata	Himalayan poplar
Rhus cotinus	Smoke tree, Indian Sumach
Pterocarpus dalbergroides	Andaman red wood
P. indicus	Malaya Padauk
P. marsupium	Indian kino tree
Canarium aphyllum	–
Ailanthus kurzii	–
A. excelsa	Maharuk tree

The Indian region has approximately world's half of the aquatic flowering plants. The aquatic families in the Indian flora are Alismataceae (5 genera, 8 spp.), Aponogetonaceae (6 spp.), Azollaceae (1 sp.), Barclayaceae (2 spp.), Butomaceae (1 sp.), Cabombaceae (2 genera, 2 spp.), Callitrichaceae (2 spp.), Ceratophyllaceae (3 spp.), Hydrocharitaceae (8 genera, 13 spp.), Isoetaceae (10 spp.), Lemnaceae (4 genera, 14 spp.), Marsileaceae (10 spp.), Najadaceae (7 spp.), Nelumbonaceae (1 sp.), Nymphaeaceae (2 genera, 7 spp.), Podostemaceae (11 genera, 24 spp.), Pontederiaceae (2 genera, 3 spp.), Potamogetonaceae (6 spp.), Ruppiaceae (1 sp.), Salviniaceae (3 spp.), Trapaceae (2 spp.), Typhaceae (4 spp.), Zanichelliaceae (1 sp.).

The families having characteristic insectivorous plants are Droseraceae (3 spp.), Nepenthaceae (1 sp.), and Lentibulariaceae (36 spp.). The parasitic families are represented by Loranthaceae (46 spp.), Santalaceae (10 spp.), Balanophoraceae (6 spp.),

Rafflesiaceae (1 sp.), Cuscutaceae (12 spp.) and Orobanchaceae (54 spp.). The species of the families Podostemaceae and Tristichaceae grow on rocks in swift flowing mountain streams. Metal tolerant species are found in the families Caryophyllacee, Cerotophyllaceae, Portulaceae, Tamaricaceae, Salvadoraceae, Thymelaeceae and Fabaceae while species of the families Chenopodiaceae, Basellaceae, Amaranthaceae and Phytolaccaceae are salt tolerant.

The size, shape, biology and economic aspects of different taxa again provide a very varied spectrum. Several highly reputed medicinal plants constitute the natural components of Indian flora. These are : *Dioscorea deltoidea* Wall., *D. prazeri* Prain and Burkill, *Atropa acuminata* Royle, *Aconitum heterophyllum* Wall. ex Royle, *A. ferox* Wall. ex Ser., *A. chasmanthum* Stapf, *Ephedra gerardiana* Wall. ex Stapf, *Nardostachys grandiflora* DC., *Rauvolfia serpentina* Benth. ex Kurz, *Saussurea sacra* Edgew., *S. lappa* Cl., *Coptis teeta* Wall., *Chlorophytum arundinaceum* Baker, *Gentiana kurroo* Royle, *Mesua ferrea* L., *Swertia chirayita* (Roxb. ex Fleming) Karsten., *Orchis latifolia* L., *Urgenia indica* Kunth, *Colchicum luteum* Baker, *Podophyllum hexandrum* Royle ex Camb., *Thymus serphyllum* L., *Curculigo orchioides* Gaertn., *Angelica glauca* Edgew., *Prangos pabularia* Lindl., *Artemisia maritima* L., *Onosma bracteatum* Wall., *Carum bulbocastanum* W. Koch, *Valeriana jatamansi* Jones, *Hydnocarpus kurzii* (King) Warb., *Datura* spp., *Holarrhena antidysenterica* R.Br., *Ferula jaeschkeana* Vatke, *Salvia moorcroftiana* Wall. ex Benth. and *Berberis* spp. There is, however, successive losses in population numbers and also changes in distribution range over a period of time. It is now well realised that some of these medicinal plants are highly endangered.

Several plants, *e.g.* species of *Arenaria, Thylacospermum, Acantholimon, Festuca* and *Juniperus* that occur especially in the high alpine meadows survive the extreme adverse ecological conditions by special adaptations The woolly species of *Saussurea* form an interesting plant of the West Himalaya. *Sapria himalayana* Griff. and *Mitrastemon yamamotoi* Makino of the family Rafflesiaceae, with only the flower (representing the whole plant) projecting from the roots of the host plant form an unusual case of botanical interest. Similarly *Balanophora dioica* R.Br., *Boschiniaekia himalaica* Hk.f. & Th. *Aeginitia indica* Roxb. are other root parasites of great morphological interest. Genera like *Galeola, Epipogium* and *Monotropa* are the best examples of saprophytes. Among the insectivorous plants *Nepenthes khasiana* Hk.f., *Drosera burmannii* Vahl, *D. peltata* Sm., *Utricularia* spp. and *Aldrovanda* spp. are of significant interest to the students of botany. *Poa litorosa* (Poaceae), the Angiosperm with highest chromosome number, also occurs in India.

The presence of a large number of primitive flowering plants in India renders the region 'a cradle of flowering plants' (Takhtajan, 1969). As many as 131 species are stated to be primitive. Some important ancient angiosperm genera are listed in Table 1.8.

Table 1.8 : Primitive Angiosperm genera in India

Families	Genera
Magnoliaceae	*Magnolia, Manglietia, Michelia, Pachylarnax, Paramichelia, Talauma*
Tetracentraceae	*Tetracentron (T. sinense var. himalayana)*
Annonaceae	*Alphonsea, Annona, Artabotrys, Cyathocalyx, Desmos, Fissistigma, Friesodielsia, Goniothalamus, Melodorum, Miliusa, Mitrephora, Orophea, Polyalthia, Trivalvaria, Unona, Uvaria*
Myristicaceae	*Horsfieldia, Knema, Myristica*
Schisandraceae	*Kadsura (K. heteroclita)*
Lauraceae	*Actinodaphne, Alseodaphne, Beilschmiedia, Cinnamomum, Cryptocarya, Dehaasia, Endiandra, Lindera, Litsea, Machilus, Neocinnamomum, Persea, Phoebe*
Chloranthaceae	*Chloranthus (C. elatior)*

It is worthwhile to appreciate the floristic diversity in some taxa/groups in India. Orchids are a well known group for their showy and long-lasting flowers and exhibit a remarkable diversity in this region. There are about 1000 species of these, with a big concentration (700 spp.) in the N.E. India. Bamboos, which are well known for their multipurpose use are also well represented. There are about 30 genera and 550 species in the world of which about 130 species occur in India. Similarly *Rhododendron*, canes, palms, ferns, Zingiberaceae members are all well represented adding to the rich floristic diversity.

Indian region is also floristically significant in having several monotypic genera such as *Alcimandra, Aspidocarya, Circaeaster, Hemsleya, Pseudodanthonia, Ischnochloa*, etc. These genera add significantly to conservation of world's genetic resources, as there are no closely related genomes of these anywhere in the world.

Table 1.9 : The number of species in some selected taxa/groups in India and in north-east India.

Genera/group	No. of species	
	India	N.E. India
Dendrobium	75	65
Bulbophyllum	50	35
Liparis	45	35
Coelogyne	35	30
Habenaria	100	45
Paphiopedilum	5	4
Vitis	70	46
Citrus	11	10
Bamboos	130	58
Rhododendron	90	80
Hedychium	40	34
Elaeocarpus	30	23
Echinocarpus	5	4
Elaeagnus	12	10
Hippophae	3	2
Ferns & Fern allies	1022	600

The flora of India shows close affinity with the flora of Indo-Malayan and Indo-Chinese region. According to Nayar (1977) 35 per cent of Indian flora has south-east Asian and Malayan, 8 per cent temperate, 1 per cent steppe, 2 per cent African, and 5 per cent Mediterranean-Iranian elements. The adventive weeds and naturalised aliens constitute only 18 per cent. This fact prompted Hooker (1904) to the erroneous conclusion that India has no flora as a separate entity but is an admixture of the floras from adjacent countries. The subsequent phytogeographers after critical analysis of flora have convincingly concluded that India has a flora of its own and as many as 5000 (33%) taxa of Indian flora are endemic to present Indian boundaries. In fact, India is said to harbour more endemic species of plants than any other region of the world except Australia. There are about 140 endemic genera distributed over 47 families. The total endemic genera in India represent 6.5% of the 2252 genera occurring in India. The largest endemic genera are *Pteracanthus* (20 spp.) and *Nilgirianthus* (20 spp.) of the family

Acanthaceae. Areas rich in endemism are the northeast India, followed by southern parts of Peninsular India and north-western Himalaya. The north-west Himalaya is estimated to harbour about 1000 endemics, but of about 3000 total species, while the Eastern Himalaya has about 1500 endemic species out of a total of about 4500 species. Peninsular India with an estimated total of 6000 species contains about 2000 endemic species. The Andaman and Nicobar Islands contribute 185 species to the endemic flora of India. The Himalaya and adjoining areas represent 2532 (50%) species followed by Peninsular India with 1788 (36%) species. The family Poaceae is one of the largest, having

Table 1.10 : The major 'hotspots' of endemic and genetic diversity of plants in India are seen in the following zones.

Hot spots	*Biogeographic zone*
Karakoram & Ladakh	Trans Himalaya
Kumaon-Garhwal Himalaya	W. Himalaya
Siwaliks	– do –
Sikkim Himalaya	E. Himalaya
Arunachal Pradesh	– do –
Lushai hills	N.E. India
Tura, Balphakram, Khasi Hills (Meghalaya)	– do –
Aravallis	Semiarid zone
Bundelkhand	Gangetic plain
Chotanagpur plateau	Deccan
Panchamarhi-Satpura ranges	– do –
Simlipal & Jeypore hills of Orissa	– do – (E. Ghats)
Bastar & Koraput hills	Deccan
Vishakhapatnam hills & Araku Valley	– do –
Tirupati-Cuddppa hills	E. Ghats
Marathwada hills	Deccan
Saurashtra-Kutch	– do –
Mahabaleshwar-Khandala ranges	W. Ghats
Agumbe-Phonda ranges	– do –
Ratnagiri & Kolaba ranges	– do –
Nilgris	– do –
Silent Valley & Wynaad	– do –
Anaamalai	– do –
Idduki-Sabarigiri	– do –
Kalakad & Agastaimalai hills	– do –

about 360 endemic taxa (Jain, 1986). The factors seem to have contributed to the high rate of endemism are (a) the barrier of high mountain region in the north, (b) separation of the southern region of the country by large water-mass of Arabian sea, Bay of Bengal and Indian Ocean, (c) the extremely arid condition in the western parts of the country and (d) humid tropical conditions in the Peninsular region and also in north-eastern region, resulting in the specification in several genera adding to the species which have continued to remain endemic. The endemism can provide clues to the mechanism of evolution through isolation and speciation. Thus we may get biological sequence in diversification of species and reasons of the restricted distribution.

Chapter 2
PHYTO SOCIOLOGICAL REGIONS OF INDIA

Considering the vast extension of the country and the variation in vegetation pattern in different areas and variation within the same area, a discussion on the floristic diversity and vegetation of the country as a whole would be too general. Several phytogeographers (Hooker, 1854, 1904; Clarke, 1898; Chatterjee, 1940, 1962; Razi, 1955; Rao, 1974; Puri, 1960; Rodgers and Panwar, 1988) have variously divided the Indian region under several phytogeographical provinces, based on the floristic composition, the naturalness of the flora and the local climate. A slightly modified version of the biogeographic classification proposed by Rodgers and Panwar (1988) has been followed in this work. Broadly, the vegetation types and floristic diversity occurring in the following 12 biogeographic zones are dealt. In addition, an account of the wetland vegetation and weeds and aliens is also disucssed.

1. Trans-Himalaya
2. West Himalaya
3. East Himalaya
4. North-East India
5. The Indian Desert
6. Semi-arid zone
7. Gangetic plain
8. Western Ghats
9. Deccan peninsula
10. Indian coasts
11. Andaman and Nicobar Islands
12. Lakshadweep Islands

It may be noted that the Himalayan region has been separated from the Trans-Himalayan belt which consists chiefly the flora of cold desert type, typical of Tibet. The vegetation of this zone is remarkably distinct from that of the remaining parts of the country. The Himalayan belt more or less has been divided further into the Western Himalaya, the Central Himalaya and the Eastern Himalaya. Since the Central Himalayan region covers exclusively the Kingdom of Nepal, this biogeographic zone has been omitted from the present discussion. The Western Himalaya according to some is again divisible into North-west Himalaya comprising parts of the States of Jammu and Kashmir and Himachal Pradesh and the Western Himalaya comprising the section of Himalaya between the river Sutlej and river Kali valley on the Indo-Nepal border (Rau, 1974). Thus, there is some overlapping of the terminology as far as the Western Himalaya and North-west Himalaya are concerned. In this account I have preferred to consider the Western HImalaya in a wider sense comprising the States of Jammu and Kashmir, Himachal Pradesh and hills of Uttar Pradesh following Clarke (1876), Chatterjee (1962) and a few others.

Chapter 3

Phyto Sociological Region of the Trans-Himalaya

The Trans-Himalaya forming the "Cold Desert of India', is the most distinct region covering the entire Ladakh district of Jammu and Kashmir and the Lahul Spiti districts of the Himachal Pradesh. The area is characterised by great extremes of heat and cold coupled with extreme dryness. The diurnal fluctuation in the temperature is very great, the noon temperature is sometimes several degrees hotter than any place in India, while during night the temperature falls many degrees below freezing point. Due to the great fluctuation in temperature, the mountains get cracked and crumbled, obviously making the region a land of "Rock ruin".

The rain fall and snow fall is also meagre. The summer is very brief with the average temperature varying from 20°-30°C and the winter temperature goes below –40°C. In certain places like Drass which is the second coldest place in the world (after Siberia) the winter temperature goes to –75°C.

This region is the most elevated zone of the earth (2900-5900 m). The topography is rough, rugged, rocky with a number of mountain peaks and sandy, extremely barren desert valleys.

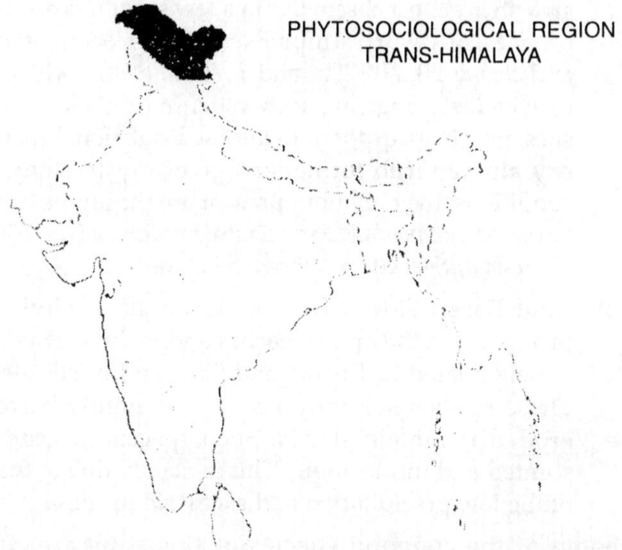

Floristic diversity

The vegetation of the area has been studied by Steward (1916-17) and recently by Kachroo *et. al.* (1977). The flora is mainly of scrub type classified as alpine scrubs. The trees are conspicuously absent except for the planted ones in the valleys along banks of rivers. Mainly (a) alpine, (b) desert and (c) oasitic elements are the characteristic feature of the cold desert flora.

(a) **Alpine Flora :** The alpine vegetation is mostly confined to narrow belts which receive part of the melting snow water and also to the upper beds of the mountain streams. The common alpine species are more or less the same as those of the Kashmir valley. Such species are *Podophyllum hexandrum* Royle ex Camb., *Lavatera kashmiriana* Camb., *Lotus corniculatus* L., *Delphinium cashmerianum* Royle, *Corydalis diphylla* L., *Astragalus rhizanthus* Royle ex Benth., *Poa annua* L., *Polygonum affine* D. Don, *Potentilla bifurca* L., *Thalaspi alpestre* L., *Oxytropis cachemeriana* Camb., *Parnassia palustris* L., *Geranium Pratens* L., *Stachys sericea* Wall., *Leontopodium* spp., *Impatiens glandulifera* Royle, *Anaphalis nubigena* DC., *Pedicularis longiflora* Rudolph and *Galium boreale* L. The alpine vegetation is very much restricted. Tree growth is almost absent but in a few sheltered places small patches of dwarf juniper scrub may occur. *Juniperus wallichiana* Hk.f. & Th. and *J. communis* L. which rarely reach a meter high are seen in dense patches on very dry sites, which are exposed to intense insulation. Where these dry alpine scrub formations give way to stony desert conditions, the only flora present are the alpine herbs like *Saxifraga pulvinaria* H.Sm., *Draba gracillima* Hk.f. & Th. and *Kobresia duthiei* Cl., all above 5500 m.

(b) **Cold Desert Flora :** The cold desert flora which is very prominent at the upper reaches and in the valleys is most closely related to Tibetan and Siberian floristic elements. These species normally occupy the highly barren and eroded mountain slopes. Such species in general are stunted and tuft forming. This is mainly due to the result of the low precipitation and excessive dryness.

Some of the common species of this zone are *Arenaria serpyllifolia* L., *Stachys tibetica* Vatck., *Nepeta tibetica* Benth., *Waldhemia stoliczkai* (Cl.) Ostenf., *Tenacetum fruiticulosum* Ledeb., *T. tibeticum* Hk.f & Th. ex Cl., *Astragalus* spp., *Chenopodium botrys* L., *Potentilla bifurca* L., *Dianthus anatolicus* Boiss., *Artemisia stricta* Edgew., *Heracleum pinnatum* Cl., *Salsola collina* Pall., *Saussurea subulata* Cl., *Hippuris vulgaris* L. and several others. In certain places on hill slopes only *Caragana versicolor* Benth., *Circium arvense* (L.) Scop., *C. wallichii* DC., *Ephedra gerardiana* Wall. ex Royle, *Acantholimon lycopodioides* (Girard) Boiss., *Chrysanthemum tibeticum* Hk.f. & Th. ex Cl., *Myricaria elegans* Royle, *Crepis stoliczkai* Cl., *Inula falconeri* Hk.f., *Leontopodium nanum* (Hk.f. & Th.) Hand.-Mazz., and the cushion-forming *Thylacospermum caespitosum* (Camb.) Schischk. are the interesting elements in this type of vegetation. *Cicer microphyllum* Benth. (wild gram) is an important genetic resource.

The cold desert plants have developed certain remarkable survival strategies in order to overcome the conditions of the cold deserts. These are :

(i) Cushion-forming habit

Cushion, clump or mat-forming habit is very common in cold desert plants. Such plants are perennial, short and sturdy with woody stem, and deep root system capable of penetrating rock crevices and fissures to provide firm anchorage (against strong wind current) and nutrition to the plant. The heavily lignified stem gives out numerous short, creeping branches just above the ground level which get repeatedly branched and densely packed with leaves and flowers forming the dense hemispheric cushion habit. Such a habit protects from the strong wind action, strong thermal radiations, drying effect of air, loss of water through transpiration, in maintaining the balance of temperature fluctuations between air and soil and from the continuous pressure of snow layer which may be of several feet thick for many months.

(ii) Diminutive or miniature habit

Although the cold desert plants are generally dwarf and stunted, some of them are so significantly reduced, one may not even notice them in fields. Species like *Lancea tibetica* Hk.f. & Th., *Lomatogonium brachyantherum* (Cl.) Fernald, *Gentianella thomsonii* Bhatt. & Sunita, *G. aquatica* L. and *Taraxacum bicolour* (Turaz.) DC., *Halerpestes tricuspis* (Max.) Hand-Mazz., *H. involucratus* Max. are often barely 1-2 cm tall with a solitary flower. In contrast to the aerial portion, the underground tap root in these species may be as long as 30 cm. The genus *Saussurea* may also be cited as an example of diminutive plants of cold desert. Several species of the genus range from hardly 2-10 cm in height. In case of *Astragalus heydei* Baker, *Corydalis crassissima* Camb., *Hedinia tibetica* (Th.) Ostenf., *Pegaeophyton scapiflorum* (Hk.f & Th.) Mary and Shaw, *Thermopsis inflata* Camb., *Microula tibetica* Benth. and *Dracocephalum heterophyllum* Benth. the plants develop deep penetrating permanent rootstocks from which annual branches are produced down among the rocks or stones and they bear leaves and flowers in clusters just above the stones or rocks.

(iii) Bushy habit

The number of woody plants in cold arid region is exceedingly low. *Caragana versicolor* (Wall.) Benth., *Ephedra gerardiana* Wall. ex Royle, *Hippophae rhamnoides* L., *Myricaria rosea* Sm. and *Lonicera hispida* Pall. ex Willd. form dense bushy habit with woody branches barely attaining 30-50 cm.

(c) **Oasitic Flora :** The Oasitic elements comprise a variety of exotic as well as some indigenous plants mainly growing along the water courses and temporary ponds near villages. Some of the common ones in this category are *Myricaria elegans* Royle, *Phragmites karka* (Retz.) Trin. ex Steud., *Equisetum* spp., *Parnassia palustris* L., *Geranium pratens* L., *Origanum vulgare* L., *Gentiana decumbense* L., etc. The introduced tree species near the villages belong to Walnut, Mulberries, Apricots and Willows all planted for their multipurpose uses. Several species of *Salix* and *Populus* are grown by the Forest Department mainly for fodder and fuel requirements. Small growth of the native *Juniperus macropoda* Boiss. is also found in some places.

A careful vegetational analysis of the zone reveals that this zone is inseparable both botanically and geologically from Tibet, both forming the Central Asia. The flora in general, therefore, have a common distributional range. About 50 per cent of the species of this zone also occur in other countries like Afghanistan, Tibet, Turkistan, Siberia, Europe and Eastern U.S.A. As many as 140 species of Ladakh are also reported from as far as Eastern United States of America. The appended tables would throw some light on the phytogeographical pattern of this region.

The Trans-Himalayan zone is a major genetic resource centre of several life-support species such as *Hordeum turkestanicum* Nevski, *Elymus dasystachys* Trin., *E. nutans* Griseb, *Rosa eglanteria* L., *R. webbiana* Wall. ex Royle (Sebeamitok), *Potentilla fruiticosa* L. (Sonmayaspa), *Sedum magae* Hamet, *Rhodiola tibetica* (Hk.f. & Th.) Fee, *Carum bulbocastanum* Koch, *C. carvi* L. (Jeera), *Allium platyspathum* Schrank, *Cicer mirophyllum* Benth. (wild gram), *Berberis petiolaris* Wall. ex G. Don, *Rubus ellipticus* Sm., *Elaeagnus angustifolia* L. and *Chenopodium* sp. There is ample scope for the development of these crops for the welfare of the local people. Genetic diversity also occurs in numerous medicinal plants in the region and to mention a few – *Aconitum heterophyllum* Wall. ex. Royle, *A. gammiei Stapf* (Atis, Buans Ponkar), *A. nepalles* L. (Vish Bougnak), *Allium humile* Kunth, *A. schoenopresum* L. (Skiche), *Arabis recta* Mill. (Solo), *A. tibetica* Hk.f. & Th. (Parpata), *Artemisa maritima* L., (Khamba), *Astragalus munroi* Benth. ex Bunge, *A. rhizanthus* Royle, *Berberis ulicina* Hk.f. & Th. (Kiraring), *B. vulgaris* L. (Kirsing, Skerpa), *Betula utilis* G.Don (Tronga), *Carum bulbocastanum* Koch (Kalajeera), *C. carvi* L. (Kongayaut), *Caragana cuneata* Baker (Taskya), *Centaurea depressa*

* Local names in brackets refers to Ladakhi names.

Bieb. (Prsaka), *Chaerophyllum villosum* Wall. ex DC., (Kalajeera), *Chenopodium botrys* L. (Sgyum, karpo), *Colchicum luteum* Baker (Kapichum), *Corydalis flebellata* Edgew., *C. tibetica* Hk. f. (Tongil), *Dactylorhiza hatagirea* (D.Don) Soo (Banglakh), *Delphinium cashmerianum* Royle (Tichulome), *Hedinia tibetica* (Th.) Ostentf. (Snankhapa), *Clematis acutifolia* Hk. f. & Th. (Emong), *Gentiana kurroo* Royle, *Gentianella moorcroftiana* (Wall. ex G.Don) Airy Shaw, Ferula *jaeschkeana* Vatke (Tnukn), *Heracleum candicans* Wall. ex DC. (Malach), *Hyoscyamus niger* L. (Langtang, Langthagcha), *Hippophae rhamnoides* L. (Terubu), *H. salicifolia* D.Don (Asmi), *Inula racemosa* Hk. f. (Manu), *Juniperus communis* L., *J. recurva* Ham. (Sukapa, Dhoop), *Lactuca tataraica* Mey., (Chimati), *Meconopsis aculeata* Royle (Chairingum), *Mentha sylvestris* L., *M. arvensis* L. (Chhiruk), *Myricaria rosea* Sm. (Umbu), *Euphorbia thomsoniana* Boiss. (Tharnu, Hrivi), *E. tibetica* Boiss., *Nepeta longibracteata* Benth. (Diyanku), *Onosma hispidum* G.Don (Demok), *Podophyllum hexandrum* Royle ex Camb. (Rodhadri, Dremmukusu), *Cachrys pabularia* (Lindl.) Herrnst. & Heyn. (Prongos), *Physochlaina praelta* Miers (Langtan, Datura), *Picrorhiza kurrooa* Royle (Hauglang), *Peganum hermala* L. (Harmal), *Plantago major* L.; *P. depressa* Willd. (Naram, Smaran), *Pedicularis longiflora* Rudolph, *P. bicornuta* Koltz. (Lugurucarpo), *Papaver nudicaule* L. (Maithuk, Sarchhing), *Rhododendron anthopogon* D.Don, *R. campanulatum* D.Don (Tajagseng), *Ranunculus hirtellus* Royle (Supkha, Chaichha), *Rheum emodi* Wall. (Lachu, Chucha), *Ephedra gerardiana* Wall., *E. intermedia* Schrank., *E. major* Host. (Chedum, Cheldug), *Rosa eglanteria* L., *R. webbiana* Wall. ex. Royle (Chharmang, Sebeamitok), *Swertia cordata* Wall., *S. petiolata* Royle (Gyatik), *Saussurea* spp. (Pangchi, Habbare, Spangohitaho), *Sonchus asper* (L.) Mill., *S. oleraceus* L., *Rhodiola tibetica* (Hk. f. & Th.) Fee, *R. quadrifida* (Pall.) Fisch. & Mey., *Thalictrum minus* L. (Chaugu), *Tanacetum tibeticum* Cl., *T. gracile* Hk. f. & Th. (Purnak, Chankar), *Taraxacum officinale* Wigg. (Rasuk, Kharchu), *Tribulus terrestris* L. (Jama), *Thymus serphyllum* L. (Tumbra), *Ulmus pumile* L. (Youmbak), *Tulipa stellata* Hk. f. (Kapi-chung).

The Trans-Himalayan region has also several unique fodder crops such as *Cachrys pabularia* (Lindl.) Herrnst. & Heyn., *Ferula jaeschkeana* Vatke, *Heracleum* spp., *Medicago* sp., *Trigonella* sp., *Potentilla fruticosa* L., etc.

Conservation of the genetic resources of these crops is highly essential as the area is now on the threshold of the development programmes.

Table 3.1 : Some cold desert species common with Afghanistan

Clematis orientalis	Ranuculaceae	*Thalictrum alpinum*	Ranuculaceae
Papaver nudicaule	Papaveraceae	*Arabidopsis thaliana*	Brassicaceae
Barbarea vulgaris	Brassicaceae	*Brassica nigra*	"
Draba alpina	"	*Mathiola flavida*	"
Sisymbrium brassiciforme	"	*Lepidium ruderale*	"
Arenaria griffithii	Caryophyllaceae	*Dianthus crinitus*	Caryophyllaceae
Silene vulgaris	"	*Stellaria palustris*	"
Geranium collinum	Geraniaceae	*Astragalus coluteocarpus*	Fabaceae
G. meeboldii	"		
Melilotus officinalis	Fabaceae	*Indigofera heterantha*	"
Potentilla atrosanguinea	Rosaceae	*Trigonella corniculata*	"
P. sericea	"	*Sibbaldia cuneata*	Rosaceae
Bergenia stracheyi	Saxifragaceae	*Rosa webbiana*	"
Sedum roseum	Crassulaceae	*Ribes alpestre*	Grossulariaceae
Ferula jaeschkeana	Apiaceae	*Bupleurum falcatum*	Apiaceae
Lonicera heterophylla	Caprifoliaceae	*Cachrys pabularia*	"
Dipsacus inermis	Dipsacaceae	*Gallium aparine*	Rubiaceae
var *mitis*		*Anaphalis contorta*	Asteraceae
Artemisia gmelinii	Asteraceae	*Centaurea wallichii*	"
Erigeron multiradiatus	Asteraceae	*Pseudognaphalium luteo-ablum*	Asteraceae
Senecio krascheninnikovii	"	*Campanula aristata*	Campanulaceae
Lapula barbata	Boraginaceae	*Swertia petiolata*	Gentianceae
Veronica perpusilla	Scrophulariaceae	*Anchusa ovata*	Boraginaceae
		Nepeta clarkei	Lamiaceae

Table 3.2 : Species of the Eastern United States occurring in Ladakh

Cystopteris fragilis	Athyriaceae	*Polygonum hydropiper*	Polygonaceae
Equisetum arvense	Equisetaceae	*Chenopodium hybridum*	Chenopodiaceae
Potamogeton pectinatus	Potamogeto-naceae	*Sagina saginoides*	Caryophyllaceae
Zannichellia palustris	Zannichelliaceae		
Triglochin maritima	Juncaginaceae	*Ranunculus trichophyllus*	Ranunculaceae
Milium effusum	Poaceae	*Barbarea vulgaris*	Brassicaceae
Deschampsia caespitosa	"	*Cardamine loxostemonoides*	"
Koehleria cristata	Gesneriaceae	*Braya humilis*	"
Catabrosa aquatica	Poaceae	*Rhodiola imbricata*	Crassulaceae
Poa alpina	"	*Saxifraga oppositifolia*	Saxifragaceae
P. pratensis	"	*Parnassia palustris*	"
P. nemoralis	"	*Potentilla anserina*	Rosaceae
Astragalus alpinus	Fabaceae	*P. fruticosa*	"
Eqilobium angustifolium	Onagraceae	*Festuca rubra*	Poaceae
Hippuris vulgaris	Hippuridaceae	*Eleocharis quinquiflora*	Cyperaceae
Primula farinosa	Primulaceae	*Scripus rufus*	Cyperaceae
Glaux maritima	Fabaceae	*Eleocharis palustris*	"
Limosella aquatica	Scrophulariceae	*Carex stenophylla*	"
Lemna minor	Lemnaceae	*C. orbicularis*	"
Oxyria digyna	Polygonaceae	*Veronica anaga-llisaquatica*	Scrophularia-ceae
Polygonum aviculare	"	*Utricularia minor*	Lentibulariaceae
P. viviparum	"	*Plantago erosa*	Plantaginaceae
		Galium aparine	Rubiaceae
		G. boreale	"

In addition 85 weedy species also occur.

Table 3.3 : Some species common with Siberia

Isopyrum grandiflorum	Ranunculaceae	*Ranunculus cymbalariae*	Ranunculaceae
Arabidopsis humile	Brassicaceae	*R. pulchellus*	"
Erysimum hieracifolium	"	*Torularia humilis*	Brassicaceae
Erodium stephanianum	Geraniaceae	*Geranium collinum*	Geraniaceae
Astragalus densiflorous	Fabaceae	*G. sibericum*	"
A. nivalis	"	*Medicago lapponica*	Fabaceae
A. subuliformis	"	*Oxytropis*	"
Rosa eglentaria	Rosaceae	*Rubus saxatilis*	Rosaceae
Rhodiola quadrifida	Crassulaceae	*Artemisia desertorum*	Asteraceae
Lectuca scariola	Asteraceae	*Tussilago farfara*	"
Gentiana aquatica	Gentianaceae	*Lomatogonium carinthiacum*	Gentianaceae
G. pedicellata	"	*Chenopodium foliosum*	Chenopodiaceae

Some species common with Europe

Chenopodium glaucum	Chenopodia-ceae	*Atriplex rosea*	Chenopodiaceae
C. foliosum	"	*Lamium amplexicaule*	Lamiaceae
Mentha longifolia	Lamiaceae	*Orobanchae cernua*	Orobanchaceae
Lycium ruthenicum	Solanaceae	*Cuscuta europea*	Cuscutaceae
Anchusa ovata	Boraginaceae	*Asperugo procumbens*	Boraginaceae
Primula sibirica	Primulaceae	*Campanula latifolia*	Campanulaceae
Hieracium virosum	Asteraceae	*Carduus edelbergii*	Asteraceae
Artemisia maritima	"	ssp. *lanatus*	
Bupleurum falcatum	Apiaceae	*Galium tricorne*	Rubiaceae
Ribes alpestre	Grossula-riaceae	*Epilobium roseum*	Onagraceae
Trifolium pratense	Fabaceae	*Saxifraga oppositifolia*	Saxifragaceae
		Geranium pratense	Geraniaceae

Table 3.5 : Some species common with Tibet

Oxytropis lapponica	Fabaceae	*Potentilla multifida*	Rosaceae
Plantago minima	Plantaginaceae	*Delphinium brunonianum*	Ranunculaceae
Corydalis crassifolia	Fumariaceae	*Silene madens*	Caryophyllaceae
Atriplex crassifolia	Chenopodi-aceae	*Polygonum sibiricum*	Polygonaceae
Rhodiola tibetica	Crassulaceae	*Arabis tibetica*	Brassicaceae
Braya thomsonii	Brassicaceae	*Stachys tibetica*	Lamiaceae
Nepeta tibetica	Lamiaceae	*Elsholtzia densa*	"
Onosma hispidum	Boraginaceae	*Mattiastrum tibeticum*	Boraginaceae
Arnebia guttata var *guttata*	"	*Lomatogonium thomsonii*	Gentianaceae
Primula elliptica	Primulaceae	*Waldheimia stoliczkai*	Asteraceae
Tanacetum tibeticum	Asteraceae	*Saussurea subulata*	"
Leontopodium nanum	"	*Inula falconeri*	"
Chrysanthemum tibeticum	"	*Rubia tibetica*	Rubiaceae
Hippuris vulgaris	Hippuridaceae	*Astragalus zanskarensis*	Fabaceae

Chapter 4

Phyto Sociological Region of the West Himalaya

The West Himalayan phyto sociological region consists of two separate phytogeographical regions, *i.e.* (a) The Western Himalaya or the Kumaon Himalaya covering mainly the Garhwal and Kumaon, and (b) the N.W. Himalaya comprising the State of Himachal Pradesh and N.W. part of Jammu & Kashmir State.

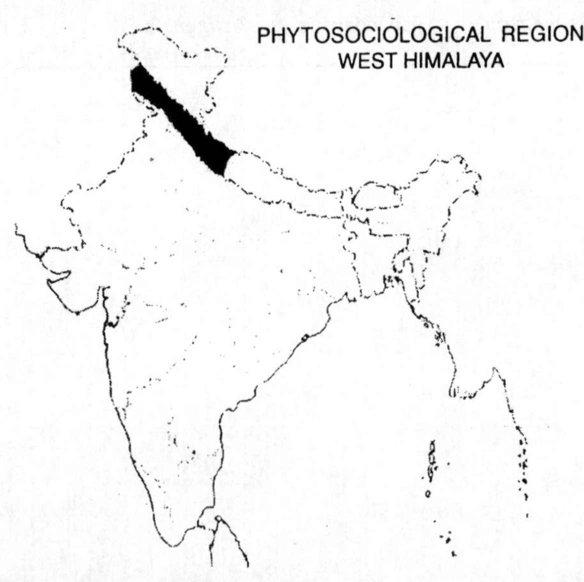

Although there are certain recognisable differences between these two regions the vegetational types and species composition more or less follows a uniform pattern, except that the N.W. Himalaya is comparatively more drier. In the present account these two regions are dealt together under the West Himalayan biogeographic zone.

Table 4.1 : Species exclusively confined to the cold arid regions

Actinocarya tibetica	Boraginaceae	*Artemisia minor*	Asteraceae
Eritrichium spathulatum var. *thomsoni*	"	*A. gmelini*	"
Mattiastrum tibeticum	"	*Acantholimon lycopodioides*	Plumbaginaceae
Tanacetum artemisioides	Asteraceae	*Senecio tibeticus*	Asteraceae
Olgaea thomsoni	"	*Leontopodium nanum*	"
Crepis multicaulis	"	*Sedum crassipes*	Crassulaceae
Stellaria tibetica	Caryophyll-aceae	*Ermania albiflora*	Brassicaceae
Astragalus maxwelli	Fabaceae	*Hedinia tibetica*	"
A. gracilipes	"	*Atelanthera perpusilla*	"
A. melanostachys	"	*Hippophae tibetica*	Elaeagnaceae
A. oxyodon	"	*Dianthus deltoides*	Caryophyllaceae
A. tribulifolius	"	*Corydalis adiantifolia*	Fumariaceae
Braya tibetica	Brassicaceae	*C. tibetica*	"
		Thylacospermum caespitosum	Caryophyllaceae

Some noticeable differences can be marked between the Western Himalaya and the Eastern Himalaya. The Eastern Himalaya is more evenly humid than the Western Himalaya. Precipitation is much higher in Eastern Himalaya than in the Western Himalaya; the timber line in the Eastern Himalaya appears around 4600 m compared to 3800 m in the Western Himalaya. The Eastern Himalaya is far richer in number of species and endemics than the Western Himalaya.

The Western Himalaya is characterised by drought resistant/cold loving plants in the landscape dominated by vast and gregarious conifer forests of 'Chir', 'Blue pine', 'Deodar' and 'Fir'. But in Eastern Himalaya except for fir (*Abies densa*) conifer forests are not vast, though species wise they are richer. Communities of oaks, laurels, rhododendrons and magnolias form the predominant vegetation in East Himalaya.

Floristic diversity

Our present knowledge of the vegetation and flora of Western Himalaya is mainly due to the efforts of numerous botanical explorers, mountaineers, surveyors, (Strachey & Winterbottom, 1846-49), foresters and even engineers during the past 150 years (Duthie, 1906; Osmoston, 1927; Stewart, 1917; Rau, 1974, 1975; Rao, 1978; Gupta, 1957, 1962). As in the Eastern Himalaya the mountains here exhibit a wide altitudinal range between the tropical foot hills to the highest limits of alpine vegetation (5000-5500 m). As the climate, topography and even edaphic conditions so greatly vary within the same region, several kinds of vegetational complexes are observed. Broadly (a) tropical (below 1000 m); subtropical (1000-2000 m); temperate (1800-3500 m); subalpine (3500-4000 m) and alpine (4000-5500 m) types of vegetations can be observed. But depending upon the local situation, slopes, edaphic condition and climate there may be some overlapping of these forest types. Several subtypes, particularly under the tropical vegetation, have been recognised and different authors have used different terms for these sub types. In brief, the following types of tropical forests are recognised in the Western Himalaya.

Tropical and Subtropical Vegetation

(i) Moist mixed deciduous Sal forests

The moist mixed deciduous Sal forest predominantly of *Shorea robusta* Gaertn. is seen in Siwaliks and on the slopes of lesser Himalaya up to an altitude of 1000 m. Also known as 'Moist Siwalik Sal forest' because of the dominance of Sal. Several other species like *Anogeissus latifolia* (Roxb. ex DC.) Bedd., *Terminalia alata Heyne* ex Roth, *Dendrocalamus strictus* Nees, also occur as main associates. Some shrubby species such as *Murraya koenigii* Spreng.,

Adhatoda zeylanica Medik are also quite conspicuous. Sal forests are absent in Kashmir.

(ii) Swamp/marshy vegetation

In the Siwalik region and Doon valley there are some patches of fresh water swamp vegetation, perhaps a relict vegetation of once a conspicuous terai belt stretching upto Assam. *Bischofia javanica* Bl., *Salix tetrasperma* Roxb., *Carallia brachiata* (Lour.) Merr., *Diospyros malabarica* (Desr.) Kostel, *Tropidia pedunculata* Bl., *Trewia nudiflora* L., *Alstonia scholaris* (L.) R.Br., *Litsea monopetala* Pers. and *Calamus* sp. (Canes) are the predominant species. These remnant patches of swamps are highly fragile and endangered. It may be interesting to note that *Carallia brachiata* (Lour.) Merr., *Tropidia pedunculata* Bl. and *Diospyros malabarica* (Dasr.) Kostel are exclusively confined to the swamp forests in W. Himalaya.

(iii) Mixed deciduous forests

This forest ascends upto 1500 m along exposed slopes with poor soil, mainly derived from shales and slates. The main components of this forest type are *Albizia procera* Benth., *Haldina cordifolia* (Roxb.) Ridsd., *Terminalia alata* Heyne ex Roth, *Dalbergia sissoo* Roxb. ex DC., *Bombax ceiba* L., *Grewia optiva* Drumm. ex Burret, *Mallotus philippensis* Muell.-Arg. and in some places *Toona ciliata* Roem. and *Bauhinia variegata* L. Another remarkable tree species in this zone is the *Ougeinia oojeinensis* (Roxb.) Hochr. which are unevenly distributed. Along the river banks, though isolated, one species consistently seen is the fast growing *Holoptelea integrifolia* (Roxb.) Planch. Often dense bamboo brakes of *Dendrocalamus strictus* Nees also occur intermixed in these forests, particularly on dry hill sides and sometimes on alluvium. The only conspicuous liana in this forest is *Bauhinia vahlii* Wt. & Arn., often masking the tree crowns and giving a gracious look when in bloom.

Although there is no clearcut dominance of any single species along the river banks, particularly in Garhwal and Kumaon region, association of *Dalbergia sissoo* Roxb. ex DC., *Acacia catechu* Willd. and *Holoptelia integrifolia* (Roxb.) Planch. can be observed. In comparatively low, dry exposed areas *Woodfordia fruticosa Kurz,* and *Ziziphus* spp. show dominance over other species.

The extremely arid north-western sector of Kashmir and Himachal Pradesh are characterised by semi-desert type of vegetation, mainly composed of *Capparis decidua* (Forssk.) Edgew., *C. spinosa* Lam., *Calotropis* sp., *Acacia* spp., *Ziziphus* spp., *Flacourtia indica* (Burm.f.) Merr., *Carissa spinarum* A. DC. and grasses of the genera *Cymbopogon, Enneapogon* and *Stipa*. In the Beas valley of Himachal Pradesh the tropical and to some extent the subtropical zone is mainly covered by *Olea ferruginea* Royle, *Punica granatum* L., *Zanthoxylum armatum* DC., apart from the common *Pinus roxburghii* Sargent. On the drier zone *Euphorbia royleana* Boiss. a fleshy, xerophytic species forms extensive patches.

(iv) Subtropical Pine Forest

The subtropical pine forests occur between 1000-1800 m in the entire Western Himalaya with the exception of Kashmir. Singh and Singh (1987) classify this forest under 'Low montane needle leaf forest with concentrated summer leaf drops'. *Pinus roxburghii* Sargent commonly known as 'Chir' pine is the predominant species attaining a height of 30-35 m. Although this species occur in its pure formation, occasionally trees like *Terminalia alata Heyne* ex Roth, *Rhododendron arboreum* Sm. and *Quercus leucotrichophora* A. Camus are seen interspersed, particularly at their upper limits, while the lower altitudinal limits are covered by Sal and other deciduous trees.

Shrub layer is poor in this forest, but grasses and some *Anaphalis* spp. are common as ground flora. Just like the *Pinus kesiya* Parl. in the Eastern Himalaya, Chir pine forests in the Western Himalaya are also not natural climax forests but long back naturalized and now a stabilized community.

A peculiar feature of this Chir pine zone is that at several places on exposed dry rocky and barren hills in Garhwal and Kumaon divisions, it gives rise to a type of subtropical scrub vegetation composed of species of *Carissa, Dodonaea, Rhus, Woodfordia, Ziziphus* and *Caesalpinia*. Extensive patches of *Euphorbia royleana* Boiss. on these dry hills is a very common sight.

The Chir pine (*Pinus roxburghii* Sargent) vegetation is absent in the Kashmir region. Here, specially on the northern exposition an *Artemisia* steppe type of vegetation occurs in which *Artemisia maritima* L. is dominant. The herbaceous members of *Eurotia* and *Kochia* of Chenopodiaceae and woody elements like *Berberis* spp., *Colutea* sp., *Sophora* and *Rosa webbiana* Wall. ex Royle, *Parrotiopsis jacquemontiana* Decne., *Indigofera heterantha* Wall. ex Brandis, *Viburnum* spp. are also prominant and sometimes this patch appears almost upto 4000 m. This type of vegetation is designated as 'Kashmir scrub' by Troll (1939).

Temperate Forests

The temperate forests are quite extensive in Western Himalaya at 1800-3000 m. They include mainly (i) Broad leaved forests, (ii) Conifer forests.

(i) Broad-leaved temperate forests

These forests are classified as 'low to mid montane Hemisclerophyllous broad leaf forests with concentrated summer leaf drop' by Singh and Singh (1987). The degree of evergreenness is greater here than in subtropical pine forests. This forest generally occupies mesic areas with annual rainfall of 1000-2500 mm and mean average temperature varying from 13°-16° C. The floristic diversity is very low. Mainly species of *Quercus* predominate.

A distinct altitudinal zonation with regard to the distribution of different species of *Quercus* in the temperate vegetation can be marked. *Q. leucotrichophora* A. Camus forest (banj oak) occupies extensive areas in lower reaches (1500-2000 m); *Q. lanata* Sm. (rianj oak) forest forms a pure stand in some pockets between 1800-2000 m. Similarly *Q. semicarpifolia* Sm. (kharsu oak) forest is predominant between 2400-3000 m.

The chief components of the oak forest are *Rhododendron arboreum* Sm., *Lyonia ovalifolia* (Wall.) Drude, *Myrica sapida* Wall., *Ilex dipyrena* Wall. and *Cornus oblonga* Wall. Occasionally Chir pine also extends into the banj oak forests. The epiphytes, particularly ferns and orchids are also rich, more so in mesic habitats as in Mandakini Valley of north Garhwal. Here, *Botrychium virginianum* Sw. grows vigorously on tree bark along with several other ferns. *Holboellia latifolia* Wall. and *Schizandra grandiflora* Hk.f. & Th. (both primitive flowering plants) are also common. The ground flora particularly in the rainy season is characterised by species belonging to *Chirita, Corallodiscus, Didymocarpus, Platystemma* of the family Gesneriaceae.

At their higher limits the oak forests again come in contact with conifers forming distinct *Quercus semicarpifolia* Sm. - *Abies pindrow* Spach. association. Although *Quercus* forests are predominant in the whole western Himalaya at 1500-3000 m, at certain isolated places such as the northern slopes near Shimla mixed forests of evergreen elements consisting of species of *Euonymus, Ilex, Litsea, Machilus* and others with bushes of *Lonicera, Rhamnus* and *Viburnum* are found.

(ii) Moist temperate broad leaved deciduous forests

Also termed 'Mis montane winter deciduous forest' by Singh and Singh (1987). These forests occupy more or less the same altitude as that of *Quercus* forests, but prefer moist hollows and depressions or as strips along hill streams. The common species are *Aesculus indica* Hiern, *Acer cappadocicum* Gleditsch, *A. caesium* Wall. ex Brandis, *Carpinus viminea* Wall. *Ulmus villosa* Brandis ex Gamble, *Juglans regia* L., *Fraxinus micrantha* Ling and *Taxus baccata* L.

In Kashmir (Dachigam Sanctuary) *Corylus-Padus* association is frequent around 2500-2900 m. In this deciduous belt other trees are *Ulmus villosa* Brandis ex Gamble, *Aesculus indica* Coleb. and *Acer cappadocicum* Gleditsch.

In many areas particularly along ravines, *Robinia pseudoacacia* L., *Salix disperma* Roxb. ex D.Don, *S. alba* L. *S. babylonica* L., *S. caprea* L. and *Morus alba* L. all introduced have become naturalized. These communities are also associated with other shrubs and trees like *Rhus succedanea* L., *Celtis caucasia* Willd.

At higher reaches of the temperate zone (2100-2900 m) specially in Garhwal, Kumaon, Himachal Pradesh and Kashmir where rainfall is below 1000 mm and where several biotic factors like continued grazing, lopping of trees, etc. operate the original forests have turned into scrub forests. Here, *Berberis jaeschkeana* Schneid., *B. pseudumbellata* Parker, *B. pachyacantha* Kohne, *Rosa webbiana* Wall. ex Royle, *R. macrophylla* Lindl., *Salix denticulata* Anders. etc. dominate over others. Champion and Seth (1968) put this type under Himalayan dry temperate forests.

(iii) Temperate conifer forests

Also termed 'Mid-montane needle-leaf evergreen forests' by Singh and Singh (1987), 'Western Himalayan moist temperate forest' by Champion and Seth (1968). These forests typically cover elevations between 2000-3000 m. The principal conifer species are *Cedrus deodara* (Roxb.) G.Don (deodar), *Abies pindrow* Spach. (silver fir), *Picea smithiana* (Wall.) Boiss. (spruce) and in limited areas *Pinus wallichiana* Jack. (blue pine). Pure forests of *Cupressus torulosa* D.Don are also quite extensive in the outer ranges mostly on lime stone rocks. Although these species form pure stands at the transitional belts, as indicated above, *Conifer-Oak-Rhododendron-Pyrus* associations become evident. At this altitude around Yamunotri and in Kumaon region *Betula-Abies* forests form a sort of transition belt between broad-leaved and conifer forests.

In the Kumaon division *Tsuga dumosa* (D.Don) Eichler forest forms the western limit of the east Himalayan mixed conifer forest. The species diversity in this forest is also considerably low.

Betula utilis D.Don - *Abies spectabilis* Spach. association around 4000 m marks the tree limit in Western Himalaya. In many places *Quercus semicarpifolia* Sm. - *Abies pindrow* Spach. also merge into this zone.

In the Kashmir valley the forests are dominated by *Pinus wallichiana* Jackson and *Abies pindrow* Spach. *Cedrus deodara* (Roxb.) G.Don is absent on the northern slopes of the Pir Panjal. But on the southern exposition around 3000-3600 m the Conifer forests are dominated by *Abies pindrow* Spach., *A. spectabilis* Spach., *Cedrus deodara* (Roxb.) G.Don, *Picea smithiana* (Wall.) Boiss. and *Taxus baccata* L. ssp. *wallichiana* Zucc. with a varying mixture of *Juniperus* spp. Occasionally patches of *Pinus gerardiana* Wall. ex Lamb is also noticed.

(iv) Subalpine vegetation

Subalpine vegetation is also termed as "High Montane Mixed Stunted Forest' by Singh and Singh (1987).

This type of vegetation appears above the tree limit, where only scattered stunted bushes of *Betula utilis* D.Don, *Juniperus communis* L., *J. wallichiana* Hk.f. & Th. come up either in pure or associated with other stunted bushes of *Rhododendron campanulatum* D.Don,

R. lepidotum Wall. ex G.Don, *R. anthopogon* D.Don, *Cotoneaster* spp., *Lonicera* spp., *Rosa* spp., *Berberis kumaonensis* Schneid., etc. The vegetation hardly grows up to 1 m high. Along the streams species of *Alectris, Caltha, Pedicularis, Potentilla, Ranunculus* with colourful flowers are common.

Apart from the above, in the alpine and subalpine zones there are excellent pastures dominated by species of *Agropyron, Brachypodium, Bromus, Poa* and *Festuca*. In these luxuriant meadows in moist localities several herbaceous perennials representing the genera *Anemone, Epilobium, Geranium, Gentiana, Polygonum* and *Bergenia* are seen. The subalpine vegetation in Kashmir appears at 3800-4100 m, where *Betula utilis* D.Don, *Abies, Juniperus, Pinus wallichiana* Jack. and *Sorbus* form the chief constituents but in stunted form. There are also 5-6 species of parasitic *Orobanche* (extending even to alpine region), of which *O. kashmirica* Cl. ex Hk.f. is endemic to Kashmir.

(v) Alpine vegetation

The Alpine vegetation marks the upper limit of vegetation in the western Himalaya. This type is met within areas between 4000-5000 m extending rarely upto 6000 m. The environmental conditions being extremely severe, only stunted scrub vegetation appears here. These high altitude grass covered alpine areas are locally known as 'Bugyals'. Many of the genera represented in the subalpine zone are also common here but in much more stunted forms. On comparatively drier sites pure *Juniperus* scrub (*J. wallichiana* Hk.f. Th. and *J. communis* L.) of less than 1 m high is a common feature. On moist localities scattered bushes of *Rhododendron anthopogon* D.Don and *Berberis* spp. are common. *Cassiope fastigiata* D.Don forms heath-like clumps and yet others present a rosette or cushion habit. Among such herbs *Paraquilegia anemonoides* (Willd.) Ulbr., *Androsace* spp., *Saxifraga* spp., *Sedum* spp. and *Bergenia stracheyi* (Hk.f. & Th.) Engl. are remarkable. The herbaceous species include apart from the genera *Polygonum, Primula, Aconitum, Orchis, Morina, Anaphalis, Inula, Thalictrum* and *Ranunculus*, several alpine rushes, sedges and grasses.

On the moraines are found peculiar members of *Saussurea*, some of them being endemic to the region. *Sausurea obovallata* Wall. popularly known as 'Brahma Kamal' is in a highly endangered state because of indiscriminate picking. The sweet scented flowers of this species are offered to the shrines of Badrinath, Kedarnath, Rudranath, and in Hemkund. The other species in this zone belong to *Caragana, Sorbus, Ephedra* and *Berberis. Picrorhiza kurrooa* Royle ex Benth., an endemic species of the alpine region is also highly endangered because of their use in herbal medicines. A cold desert element of Ladakh, *i.e. Thylacospermum caespitosum* (Camb.) Schischk. (cushion forming plant) is also present here. It may also be of interest to note that the highest altitude known for a flowering plant in western Himalaya is 6300 m, on Mt. Kamet, where a specimen of *Christolea himalayensis* Jafri of the family Brassicaceae was collected by Gurudial Singh (Rao, 1978).

The alpine flora in western Himalaya are all spectacular mostly appearing during July to August when the whole valley gets sprinkled with a variety of colourful flowers – a remarkable sight.

The famous Valley of Flowers (Bhyunder Valley) is a great botanical paradise as well as a tourist attraction.

Affinities

The study on the phytogeographical affinities of the flora of the north-west Himalaya with the surrounding regions is indeed very fascinating. The close affinity between the flora of the north-west Himalaya with those of Europe, the near East and middle East is well established (Legris, 1963; Gupta, 1962, 1964, 1982; Meusel, 1971; Rau, 1974, 1975, 1981).

The European and Central Asian elements are frequent in areas west of the river Sutlej, while the Chinese elements extend from Yunnan in the east right through the East HImalayan ranges. From the dry mountains of western and middle Asia many elements have spread to the western ranges of the Himalaya. This influx is greatly due to the arid and dry conditions prevailing here particularly in the interior ranges of Ladakh, Lahul and Spiti valley. Several such species of middle Asia like *Rosularia alpestris* (Kar. & Kir.) Boiss., *Salix karelinii* Turcz ex Stsch., *Sorbaria tomentosa* (Lindl.) Rehder, *Lathyrus humilis* Fisch. ex Spreng., *Acantholimon lycopodioides* (Girard) Boiss., *Myricaria squamosa* Desv., *Oxytropis microphylla* (Pall.) DC., *Biebersteinia odora* Steph. ex Fisch., etc. are found in the north-west Himalayan region.

Cedrus deodara (Roxb.) G. Don common on the west Himalayan slopes is also distributed to as far away as Afghanistan. The eastern limits of the distribution of the species is the western part of Nepal. Based on this as well as on the distribution pattern of several species it has been concluded that the zone of transition between the Phytogeographical regions of Eastern and Western Himalaya is approximately the area between 80° E to 84° E longitude (Stearn, 1960).

Regarding other Gymnosperms, as far as number of species are concerned, although eastern Himalaya are richer, there exist vaster coniferous forests in the Western Himalaya. *Pinus gerardiana* Wall. ex Lamb, *Juniperus macropoda* Boiss., *Picea smithiana* Boiss. are some gymnosperms distributed in the north-west Himalaya but are absent in the eastern Himalaya. Similarly *Ephedra* a genus of important medicinal herbs is well represented in the north-western Himalaya with six species while only one species occurs in the eastern Himalaya. Several species like *Larix griffithiana* Carr., *Picea spinulosa* (Griff.) Henry, *Cephalotaxus griffithii* Hk.f., *Gnetum montanum* Markgraf, *Cycas pectinata* Griff., etc. found in east Himalaya are absent beyond E. Nepal.

Several temperate species from Europe and other regions have also found their way to this region. Some of these species are *Melilotus officinalis* (L.) Lamk., *Medicago sativa* L. ssp. *falcata* (L.) Naeg. & Thell., *Aconogonum alpinum* (All.) Schur, *Trifolium repens* L., *Lotus corniculatus* L., *Onopordum acanthium* L. *Chenopodium foliosum* (Moench.) Aschers., *Centaurea iberica* Trevir ex Spreng., *Geranium pratense* L., *Mentha longifolia* (L.) Huds., *Carthamus lanatus* L., *Artemisia absinthium* L., *Briza media* L., *Dactylis glomerata* L., *Poa trivaialis* L., *Draba nemorosa* L., *Erophila verna* (L.) Besser, *Barbaraea vulgaris* R.Br., *Cardamine impatiens* L. etc.

There are also other introductions like *Datura suaveolens* Humb. & Bonpl. ex Willd., *D. stramonium* L., from Tropical America, *Nicandra physaloides* (L.) Gaertn. from Peru, *Ipomoea purpurea* (L.) Roth from C. America, *I. carnea* Jacq. from S. America, *Martynia annua* L. from America, etc. which have now become naturalized.

Viola biflora L. a common species in the northwestern Himalaya is also known from Europe, Siberia, Central Asia, North Korea, Japan, North America as well as in the Central and Eastern

Himalaya regions. Similarly *Capparis spinosa* L., is known from Afghanistan to Nepal, W. Asia, Europe. *Pos alpina* L. is another species which is widely distributed in Pakistan, India, Europe, Mediterranean region, Middle east to C. Asia and North America.

However some species like *Cotoneaster frigidum* Wall. ex Lindl., *Rubus calycinus* Wall. ex. G.Don, *R. acuminatus* Sm., *Androsace delavayi* Franchet, *Osmanthus suavis* King ex Cl., *Boschniakia himalaica* Hk.f. & Th. ex Hk. f., etc. originating in S.W. China reach only upto Kumaon in Uttar Pradesh. Similarly, there are species like *Cypripedium elegans* Reichb. f., *C. himalaicum* Rolfe, *Roscoea purpurea* Sm., *Primula tibetica* Watt, *P. primulina* (Spreng.) Hara extend from S.E. Tibet to Uttar Pradesh.

Circaeaster agrestis Max. another plant of north-western China extends across Tibet to the Himalaya as far west as Garhwal. Similarly, there are several species distributed not only in the north-western Himalaya but all along the Himalayan range upto S.E. Asia, Burma, etc.

The extraneous elements of the flora from S.W. China, C. Asia, W. Asia, Europe have mixed up with the local species during the course of evolution giving rise to the north-west Himalayan flora as we know it today. At the same time, some of these migratory elements have remained unchanged. Example of such species are *Melilotus alba* Lamk., *Melica nutans* L., *Potentilla fruiticosa* L., *Flemingia strobilifera* R.Br., *Nasturtium officinale* (L.) R.Br., etc.

Some species are also treated as related subspecies or close varients of the species found in Eurasian regions. Presumably these taxa migrated to this region during the Pleistocene glaciation and subsequently adopted to the new environs resulting in their present status.

Although the north-west Himalayan flora is an admixture of floras from Mediterranean region, C. Asia, Europe, S.W. China, etc. a careful analysis reveals that the north-west Himalayan region is also rich in endemic species (Table 4.2 and 4.3).

Table 4.2 : Some species of Chinese origin widely distributed in the Himalaya

Name	*Family*	*Distribution*
Valeriana jatamansi	Valerianaceae	Afghanistan to S.W. China, Burma.
V. hardwickii	Valerianaceae	Pakistan to S.W. China, Burma, S.E. Asia
Calotropis gigantea	Asclepiadaceae	Throughout Himalaya, S.E. Asia, China
Cariocrinum giganteum	Liliaceae	Kashmir to S.W. China, Burma
Dactylorhiza hatagirea	Orchidaceae	Pakistan to S.E. Tibet, Europe, N. Africa & S.W. Asia
Primula denticulata	Primulaceae	Afghanistan to S.E. Tibet, Burma
Taxus baccata ssp. *wallichiana*	Taxaceae	Afghanistan to S.W. China Burma, S.E. Asia
Symplocos paniculata	Symplocaceae	Pakistan to S.W. China, Burma, Japan, S.E. Asia
Jasminum dispermum	Oleaceae	Kashmir to S.W. China, S.E. Asia
Buddleja asiatica	Loganiaceae	Pakistan to Bhutan, C. & S. China, Burma, S.E. Asia
Acer oblongum	Aceraceae	Pakistan to S.W. China, Burma, S.E. Asia
Hedera nepalensis	Araliaceae	Afghanistan to S.W. China, Burma
Swida oblonga	Cornaceae	Kashmir to S.W. China, Burma, S.E. Asia
Leycesteria formosa	Caprifoliaceae	Pakistan to S.W. China, Burma
Lonicera webbiana	Caprifoliaceae	Afghanistan to S.W. China

The West Himalayan biogeographic zone is extremely rich in plant life and abounds in genetic diversity of medicinal plants, timber plants, wild relatives of crop plants and other economic species. Maximum genetic diversity can be observed in a number of timber species, both conifers and broad-leaved species.

Table 4.3 : Some endemic species of N.W. Himalaya

Name	Family	Distribution
Androsace primuloides	Primulaceae	Jammu and Kashmir, 3000-4000 m
Hedysarum cachemirianum	Papilionaceae	Jammu and Kashmir, 2500-4000 m
H. microcalyx	Papilionaceae	Jammu and Kashmir to Uttar Pradesh, 2500-4000 m
Saussurea atkinsonii	Asteraceae	Jammu and Kashmir to Uttar Pradesh, 3000-4500 m
S. clarkei	Asteraceae	Jammu and Kashmir, *ca* 4400 m
Poa falconeri	Poaceae	Jammu and Kashmir, *ca* 4000 m
P. koelzii	Poaceae	Jammu and Kashmir, *ca* 5000 m
Puccinellia stapfiana	Poaceae	Jammu and Kashmir, *ca* 5000 m
P. thomsonii	Poaceae	Jammu and Kashmir, *ca* 5000 m
Catabrosella himalaica	Poaceae	Jammu and Kashmir, *ca* 4500 m
Arabis tenuirostris	Brassicaceae	Jammu and Kashmir, *ca* 3000 m
Hyalopoa nutans	Poaceae	Jammu and Kashmir, *ca* 3500-4500 m
Delphinium roylei	Ranunculaceae	Jammu and Kashmir, Himachal Pradesh, 1600-2500 m
Carex munroi	Cyperaceae	Himachal Pradesh, *ca* 3800 m
Microschoenus duthiei	Cyperaceae	Uttar Pradesh, *ca* 5300 m
Dicranostigma lactucoides	Papaveraceae	Uttar Pradesh, Nepal, 2700-4000 m
Draba lasiophylla	Brassicaceae	Jammu and Kashmir, Himachal Pradesh, 3800-4000 m
Erophila teperrima	Brassicaceae	Jammu and Kashmir, *ca* 4200 m
Christolea acaposa	Brassicaceae	Jammu and Kashmir, *ca* 4950 m

A systematic evaluation and assessment of genetic variability among different populations of most of the species in the Himalaya remains a challenge to botanists. Important conifer species that cover vast areas in West Himalaya are – *Abies pindrow* Royle 'morinda', *Cedrus deodara* (Roxb.) G.Don 'deodar', *Picea smithiana*

(Wall.) Boiss. 'rai', *Pinus roxburghii* Sargent 'chir', *Pinus wallichiana* Jack. 'kail'.

Among the broad-leaved species, genetic diversity and variability can be observed in the following species : *Acacia catechu* (L.f.) Willd. 'khair', *Acer* spp. (*A. oblongum* Wall. ex DC., *A. caesium* Wall. ex Brandis, *A. laevigatum* Wall., *A. cappadocicum* Gleditsch., *Haldina cordifolia* (Roxb.) Ridsd., 'haldu', *Aesculus indica* (Colebr. ex Camb.) Hk. 'pangar', *Albizia odoratissima* (L.f.) Benth. 'kalasiris', *A. julibrissin* Duraz. 'bhondir', *A. chinensis* (Osbeck) Merr. 'siran', *Anogeissus latifolia* (Roxb. ex DC.) Wall. ex Bedd. 'bankli', *Bauhinia purpurea* L. 'kachnar', *Betula alnoides* Buch.-Ham. ex D. Don 'keth bhunj' *Bischofia javanica* Bl. 'bhillar', *Bombax ceiba* L. 'semal', *Buchanania lanzan* Spreng. 'kath bhilanda', *Carpinus viminea* Lindl. 'chamkhrik', *Dalbergia sissoo* Roxb. ex DC. 'shisham' *Diospyros exsculpta* Buch-Ham. 'tendu'. *Ehretia acuminata* R. Br. 'puna', *Garuga pinnata* Roxb. 'kharpat', *Gmelina arborea* L. 'khambar', *Holoptelea integrifolia* Planch. 'kali papri', *Hymendodictyon excelsum* (Roxb.) Wall. 'kukarkat', *Lannea coromandelica* (Houtt.) Merr. 'jhingan', *Lagerstroemia parviflora* Roxb. 'dhaura-dhauri', *Mangifera indica* L. 'am', *Melia azedarach* L. 'dainkan', *Ougeinia oojeinensis* (Roxb.) Hochr. 'sandan', *Drypetes roxburghii* (Wall.) Hurusawa 'putijiva', *Quercus floribunda* Lindl. ex A. Camus 'moru', *Q. leucotrichophora* A. Camus 'banj', *Q. semecarpifolia* Sm. 'kharshu', *Schleichera oleosa* (Lour.) Oken 'Kusum', *Shorea robusta* Gaertn. f. 'sal', *Stereospermum chelonoides* (L.f.) DC. 'padal', *Syzygium cumini* (L.) Skeels 'jamun', *Terminalia bellirica* (Gaertn.) Roxb. 'bahera', *T. chebula* Retz. 'har', *T. alata* Heyne ex Roth 'sain', *Ulmus villosa* Brandis ex Gamble 'mai'.

The West Himalayan region is also a potential source of many essential oil yielding and temperate drug plants. The important medicinal plants are *Aconitum heterophyllum* Wall. ex Royle, *A. chasmanthum* Stapf, *A. deinorrhizum* Stapf, *Nardostachys grandiflora* DC., *Dactylorrhiza hatagirea* (D. Don) Soo, *Picrorhiza kurrooa* Royle ex Benth., *Rheum nobile* Hk. f. & Th., *R. australe* D. Don, *Swertia chirayita* (Roxb. ex Fleming) Karsten, *Bergenia ciliata* f. *ligulata* Yeo, *Ephedra gerardiana* Wall. ex Stapf, *Hyoscyamus niger* L., *Atropa acuminata* Royle, *Coptis teeta* Wall., *Podophyllum hexandrum* Royle ex Camb., *Saussurea lappa* Cl., *Gentiana kurroo* Royle, *Acorus calamus* L., *Dioscorea prezeri* Prain & Burkill, *Physochlaina* spp. and *Berberis* spp.

The West Himalayan biogeographic zone is also known for its rich genetic diversity in several wild plants of food value. Numerous wild species are used as leafy vegetables, as fruit plants and so on. Some of these are enumerated below :

Leafy vegetables

Alternanthera sessills (L.) R.Br. ex DC., *Amaranthus tricolor* L., *Cardamine hirsuta* L., *Rorippa indica* (L.) Hiern, *Cicer microphyllum* Benth., *Chenopodium album* L., *C. murale* L., *Oxalis acetosella* L., *O. corniculata* L., *Polygonum alpinum* All., *Ranunculus sceleratus* L., *Rumex acetosa* L., *R. hastatus* D. Don, *Rhodiola imbricata* Edgew., *Sonchus brachyotus* DC., *Drimia indica* (Roxb.) Jessop, *Allium* spp., etc.

The wild plants used for edible fruits are *Cornus capitata* Wall., *Ficus aruriculata* Lour., *F. hispida* L.f., *F. palmata* Forssk., *Grewia elastica* Royle, *Malus baccata* (L.) Borkh. *Morus australis* Poir., *Olea ferruginea* Royle, *Pyrus pashia* (Buch.-Ham.) D. Don, *Sorbus lanata* (D.Don) Schauer, *Berberis aristata* DC., *B. asiatica* Roxb. ex DC., *B. lycium* Royle, *Myrsine africana* L., *Prunus prostata* Labill, *Punica granatum* L., *Rosa macrophylla* Lindl., *Rubus biflorus* Buch.-Ham. ex Sm., *R. ellipticus* D.Don, *S. paniculatus* Sm., *R. fruticosus* L., *Solanum erianthum* D.Don, *S. surattense* Burm. f., *S. melongena* var. incanum (L.) Kuntze, *Fragaria vesca* L., *Potentilla fruitcosa* L., *Mukia maderaspatana* (L.) Roem., *Carissa carandas* L., *Ziziphur* spp., *Myrica esculenta* Buch.-Ham., etc. Similarly seeds of *Juglans regia* L., *Olea ferruginea* Royle, *Pinus gerardiana* Wall. ex Lamb., *Terminalia chebula* Retz., *Cassia obtusifolia* L., *Chenopodium album* L., *Amaranthus* spp., *Impatiens balsamina* L. are consumed as food.

The genetic diversity in species related to our crop plants is also extremely rich in this biogeographic zone. Wild relatives of cereals and millets such as *Avena fatua* L., *A. sterilis* L. spp. *ludoviciana* Dur., *A. barbata* Pott ex Link, *Patropyrum tauschii* (Cosson) A. Love, *Digitaria sanguinalis* Scop., *Hordeum spontaneum* C. Koch, *Critesion glaucum* (Steud.) A. Love, *C. brevisubulatum* (Trin.) A. Love spp. *turkestanicum* (Nevski) A. Love, *Elymus dahuricus* Griseb., *E. nutans Griseb.*, *Leymus secalinus* (Georgi) Tzvelev, *Eremopyrum buonepartis* Nevski, *E. distans* Nevski, *E. orientale* Taub. & Spach, *Pennisetum orientale* Rich., pulse crops (legume) like *Cicer microphyllum* Benth., *Lathyrus aphaca* L., *Flemingia procumbens* Roxb., *Mucuna capitata* Sweet ex Wt. & Arn., *Trigonella emodi* Benth., *Vigna vexillata* (L.) Rich. var. *angustifolia* (Schum. & Thonn.) Baker, *V. sublobata* (Roxb.) Babu & Sharma, *V. umbellata* (Thunb.) Ohwi & Ohashi; spices and condiments such as *Allium jacquemontii* Kunth, *A. tuberosum* Rottl. ex Spreng., *A. schoenoparsum* L., *Carum bulbocastinum* W. Koch and fruit species like *Duchesnea indica* (Andr.) Focke, *Elaeagnus angustifolia* L., *Ficus palmata* Forssk., *Morus* spp., *Prunus undulata* Buch.-Ham. ex D. Don, *P. cerasioides* D.Don, *P. cornuta* (Wall. ex Royle) Steud., *P. napaulensis* (Ser.) Steud., *P. prostrata* Labill., *P. tomentosa* Thunb., *P. amygdalus* Batsch (*P. domestica* L.) *P. pashia* Buch.-Ham. ex D.Don, *Malus baccata* (L.) Borkh., *Ribes graciale* Wall., *R. nigrum* L., *Rubus ellipticus* Sm., *R. alceifolius* Poir., *R. fruiticosus* L., *R. niveus* Thunb., *R. lanatus* Wall. ex Hk.f., *R. reticulatus* Wall. ex Hk. f., *Ziziphus mauritiana* Lam., are richly represented in this zone. Besides the above, wild relatives of sugar cane- *Saccahrum filifolium* Steud., *S. rufipilum* Steud., *S. arundinaceum* Retz. and *S. spontaneum* L. also exhibit maximum diversity in this zone.

Chapter 5

Phyto Sociological Region of the Eastern Himalaya

The limits of the Eastern Himalaya as a phyto sociological region are not well defined. While some authors include Sikkim, Darjeeling district of W. Bengal and Arunachal Pradesh under Eastern Himalaya, yet others include only Arunachal Pradesh and treat Sikkim and Darjeeling under the Central Himalaya. However,

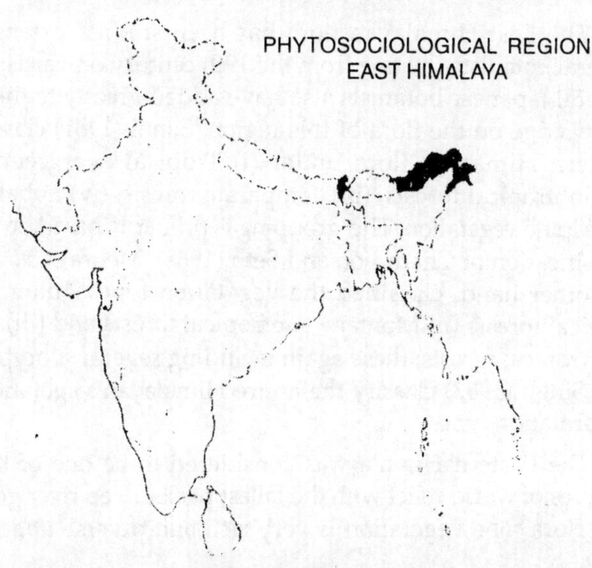

as accepted by a majority of biogeographers the author has included Sikkim, Darjeeling district of W. Bengal and Arunachal Pradesh in the eastern Himalaya.

Rodgers and Panwar (1988) include East Himalaya in their Himalayan Biogeographic zone. Although the entire Himalayan mountains form a continuous, geologically connected chain, the climate and the resultant vegetation types are so different that it is worthwhile to treat western and eastern Himalayas as distinct biogeographic zones. The stretch of Eastern Himalaya approximately runs to 850 km. Here, the Himalaya rises rather abruptly from the plain and hence the sub-Himalayan zone is not distinct compared to the western Himalaya. The Eastern Himalaya is more mesic. The high degree of precipitation is due to the abruptly rising hills that directly confront the moisture laden monsoon wind, blowing from the Bay of Bengal. The special horse shoe-shaped arrangement of the fold of the mountains of the Eastern Himalayan zone ensure plenty of rains in most of the places. Along with the region of high rainfall, there are also regions with moderate to low rainfall zones which account for a great diversity in plant wealth here. The altitude ranges from 1500 m to the lofty ice capped mountains of Kanchenjunga at the height of 8598 m.

Vegetational Diversity

The East Himalayan flora has been studied extensively by several botanists starting from the 19th century onwards. Recently several Japanese botanists also have added greatly to the existing knowledge on the flora of this region. Sahni (1981) classified the eastern Himalaya flora under (i) Tropical evergreen forests, (ii) Subtropical forests, (iii) Temperate forests, (iv) Subalpine and (v) Alpine vegetation. This grouping is primarily based on the forest classification of Champion and Seth (1968). Mehra *et. al.* (1985) on the other hand, classified the vegetation types under (i) Moist tropical forests (ii) Montane subtropical forests and (iii) Montane wet tropical forests; these again including several subtypes. Singh and Singh (1987) classify the entire Himalayan vegetation under 11 formation types.

The Eastern Himalaya is considered to be one of the major features of world relief with the tallest peaks, deep river gorges and rare flora. The vegetation is very rich and diverse abounding in spectacular flora of some of the tallest trees in India, tree ferns, orchids, primulas and blue poppies. The mountain slopes are covered by a variety of colourful rhododendrons. The region is also the habitat of many botanical curiosities and botanical rarities. The lower reaches of the Eastern Himalaya are considered to be the sanctuaries of ancient flora, as evident by the presence of several primitive flowering plants. Takhtajan (1969) treats this region as the 'Cradle of flowering plants'. The Eastern Himalaya is far more evenly humid than the Western Himalaya. The high humidity is conducive for the tree growth and therefore the timber line or the upper limit of the tree vegetation in this sector goes upto 4600 m as compared to 3800 m in Western Himalaya. Floristically this region also acts as a gateway for migration of flora from the adjacent countries like Burma, China, Japan Bhutan, etc.

The available knowledge of the vegetation types of the Eastern Himalaya is mainly by the workers like Griffith (1847) Hooker (1849), Bor (1938), Kingdon-Ward (1960) and recently by Hara (1968-71), Sahni (1969, 1979), Deb and Dutta (1971), Rao (1974) and Rao and Hajra (1986). All the above workers have tried to classify the vegetation types, essentially based on altitudinal zones. Depending upon the altitude the forests vary from tropical types to alpine vegetation, each having numerous subtypes.

1. Tropical Forests

(a) Tropical evergreen forests

Tropical evergreen forests extend from foothills upto 1000 m and are met within the Terai region of Arunachal Pradesh. According to Gamble (1875) these forests are termed as 'Lower hill forests', whereas Champion (1936) called them as 'East sub-Himalayan wet mixed forests'. The rainfall pattern varies from 260-500 cm. Depending upon the location and precipitation, these forests also habour a number of deciduous species and therefore, they may even be classified as tropical moist deciduous to semi evergreen forests. Also, the species composition varies from region to region and even district to district in Arunachal Pradesh. Therefore, it is difficult to draw any generalization regarding species composition. The species listed here therefore, may not be typical of all regions in the Eastern Himalaya.

Tropical evergreen forests exhibit distinct stratification with the top storey consisting of tall trees of *Dipterocarpus retusus* Bl., *Artocarpus chama* Buch.-Ham., *Shorea robusta* Gaertn.f. and *Tetrameles nudiflora* R.Br. ex Benn. These trees are heavily plastered with lichens and festooned with climbers and epiphytes. The other comparatively smaller tree species are *Artocarpus fraxinifolia* Wt., *Gynocardia odorata* R.Br., *Saurauia napaulensis* DC., *Syzygium cumini* (L.) Skeels, *Ailanthus integrifolia* Lamk. ssp. *calycina* (Piera) Nooteb., *Knema angustifolia* (Roxb.) Warb., *Duabanga grandiflora* (Roxb. ex DC.) Walp., *melina arborea* L. (deciduous), and *Oroxylum indicum* (L.) Vent. The tree ferns are often seen associated with *Pandanus nepalensis* St. John. The hill slopes with wild banana form the prominent feature of the vegetation. the epiphytic orchids are quite numerous, of which the genera *Dendrobium* and *Cymbidium* top the list.

In the Tirap district of Arunachal Pradesh *Dipterocarpus retusus* Bl. is mainly associated with *Shorea assamica* Dyer, *Mesua ferrea* L., *Knema angustifolia* (Roxb.) Warb. and *Terminalia myriocarpa* Heurck & Muell. Yet in other places Lauraceous members such as *Cinnamomum pauciflorum* Nees, *C. tamala* Nees, *Litsea monopetala* Pers. are dominant. Another remarkable group of herbaceous plants that exhibit maximum diversity in these forests are the members of Zingiberaceae (ginger family). As many as 18 species of *Hedychium* are reported from Arunachal Pradesh alone.

(b) Tropical grasslands

The grassland vegetation as a biotic climax makes its appearance around 1000 m, where both subtropical evergreen forests and grasslands coexist. The common grass species are *Arundinella bengalensis* (Spreng.) Druce, *Setaria palmifolia* (Willd.) Stapf, *Phragmites karka* Trin. ex Steud., *Arundo donax* L., *Imperata cylindrica* (L.) Raeus., *Erianthus longischorus* Anders., *Saccharum spontaneum* L. and among the bamboos which also form the component of the grasslands, mention can be made of *Chimonocalamus griffithianus* (Munro) Hsuch & Yi, *Cephaloschizostachym latifolium* (Munro) R. Majumdar and *Dendrocalamus hookeri* Munro. In the higher elevations of Subansiri district of Arunachal Pradesh, caespitose bamboos, *Phyllostachys bambusioides* Sieb. & Zucc. and *Dendrocalamus sikkimensis* Gamble are very common. The valuable broom grass belonging to

Thysanolaena maxima Kuntze is also seen on the hill sides as a component of grassland vegetation. At several places *Alpinia nigra* (Gaertn.) Burrt. (Zingiberaceae) grows in pure stands among these grasslands.

(c) The subtropical forests

The subtropical forests mainly occupy the elevations between 1000-2000 m, where the rainfall is between 200-400 cm. These forests are termed as Middle hill forests (Gamble, 1875); Bengal subtropical hill forests (Champion, 1936) and mixed broad leaved forests (Kanai, 1966). Grierson and Lang (1983) in their account of Flora of Bhutan have classified this type of forest under warm broad leaved forests. The forest vegetation is dominated by *Ficus-Castonopsis-Callicarpa* association in the lower reachers and by *Schima-Castonopsis-Enelhardtia* association in the higher region. The common species are *Schima wallichii* Choisy, *Castanopsis indica* A.DC., *Saurauia fasciculata* Wall., *Stereospermum colais* (Dill.) Mabberly, *Firmiana colorata* (Roxb.) R.Br., *Exbucklandia populnea* (R.Br. ex Griff.) R.Br., *Prunus cerasioides* D.Don, *Trevesia palmata* Vis., and *Wendlandia* spp. In certain places along the water courses *Alnus nepalensis* D.Don is quite extensive. On dry slopes *Rhododendron arboreum* Sm., *Schima wallichii* Choisy, *Garuga pinnata* Roxb. are also common, but nowhere abundant. Palms are also well represented here. Among the dominant ones *Livistona jenkinsiana* Griff., *L. speciosa* Kurz, *Wallichia caryotoides* Roxb., *W. disticha* T. Anders., *Calamus erectus* Roxb., *C. floribundus* Griff. are important. Among the shrubs the important ones are *Rubus ellipticus* Sm., *Artemisia nilagirica* (Cl.) Pamp., *Boehmeria macrophylla* Horn. var. *canescens* (Wedd.) Lang, *Osbeckia sikkimensis* Craib, *O. nepalensis* Hk., *Melastoma nepalensis* Lodd and *Maesa chisia* D.Don. The ground flora, mainly the herbaceous ones belong to the families Asteraceae, Solanaceae, Zingiberaceae and Urticaceae. The orchid flora is also diverse.

At comparatively drier climates *Trema orientalis* Bl. is very common and is associated with *Garuga pinnata* Roxb., *Stereospermum colais* (Dilw.) Mabberly, *Litsea monopetala* Pers., *Baccaurea ramiflora* Lour. and *Bischofia javanica* Bl. *Pinus roxburghii* Sargent (Chir pine), which appears quite extensively at this altitude in the western Himalaya is very rare in the eastern Himalaya. This species occurs rarely in Sikkim and in Kameng district of Arunachal Pradesh. *Cephalotaxus griffithii* Hk.f., *Podocarpus neriifolius* D.Don also occur in Kameng district approximately at this elevation.

2. Temperate Forests

This zone (1800-3500 m) is again distinguished under different forest types as follows :

(a) The cool broad leaved forests of eastern Himalayan wet temperate forests

Also termed as 'Mid-montane broad leaved ombrophilous forest' by Singh and Singh (1987). This forest appears upto elevation of 2500 m. The species are *Acer campbellii* Hk.f. & Th., *Corylopsis himalayana* Griff., *Ilex* spp., *Persea* spp., and *Symplocos* spp.

(b) Evergreen Oak forests

Also known as Temperate Oak forest and is noticed at 1800-2600 m altitude. The rainfall varies from 200-300 cm. In this zone mostly temperate rain forest species such as, *Lithocarpus elegans* (Bl.) Hotus ex Soepadmo, *Rhododendron arboreum* Sm., *R. lepidotum* Wall., *Litsea thomsonii* Meissn., *Styrax serrulatum* Roxb., *Acer campbelli* Hk.f. & Th., *Castanopsis purpurella* (Miq.) Balakr., *C. tribuloides* A.DC., *Juglans regia* L. and species of *Symplocos, Skimia, Randia* and *Lyonia* are common. *Magnolia campbelli* Hk.f. & Th., associated with *Rhododendron falconeri* Hk.f., *R. thomsonii* Hk.f., *Symplocos cochinchinensis* ssp. *laurina* (Retz.) Nooteb. and *Michelia doltsopa* Buch.-Ham. are very common particularly in Arunachal Pradesh.

(c) The Temperate Rhododendron-Conifer forests

This type occurs between 2500-3500 m, more predominantly in Sikkim region. The conifers such as *Cupressus torulosa* D.Don, *Tsuga dumosa* (D.Don) Eichler, *Abies delavayi* Franch. (Chinese fir) localised in Kameng district, *Larix griffithiana* Gord. (only deciduous conifer in India), *Picea morinda* Link (in Sikkim) and *P. brachytyla* Pritz. (in Kameng district) and *P. spinulosa* Henry are some of the conifer species in this zone.

These coniferous belts are also associated with broad leaved plants like *Rhododendron arboreum* Sm., *R. hodgsoni* Hk.f., *R.*

griffithianum Wt., *Acer sterculiaceum* Wall., and *Pyrus sikkimensis* Wenzing, shrubs are also numerous but the important ones are *Rosa sericea* Lindl., *Prinsepia utilis* Royle, *Rhododendron lepidotum* Wall. ex G.Don, *Agapetes serpens* Sleumer and *Lyonia ovalifolia* (Wall.) Drude. At higher altitudes only conifer-*Rhododendron* association is dominant.

3. Sub-Alpine Forests

This zone ranges between areas from 3600-4200 m altitude. The forest is also termed as Moist alpine scrubs by Champion (1936); Alpine scrub and meadows by Kanai (1966); *Juniperus-Rhododendron* scrub by Grierson and Lang (1983). The tree species in this zone are very rare, and only the dominant species is *Abies densa* Griff. associated with *Betula utilis* D.Don and rarely by *Juniperus wallichiana* Hk.f. & Th. Only the bushy and herbaceous flora is conspicuous in this zone. These species belong to *Berberis, Cotoneaster, Ribes, Rhododedron* and *Salix*. Herbaceous species of the genera *Meconopsis, Cardamine, Potentilla, Astragalus, Corydalis, Primula, Pedicularis, Polygonum, Rheum, Saussurea* and in other comparatively drier places *Cassiope* and *Anaphalis* are very common. At higher reaches *Rhododendron hodgsonii* Hk.f., with pink flowers along with *R. anthopogon* ssp. *hypenanthum* (Balf. f.) Cullen, *R. campanulatum* D.Don and *R. thomsonii* Hk.f. are very frequent. In southeast Sikkim some of these rhododendrons are associated with *Salix* spp. and many bulbous or tuberous species represented by *Codonopsis, Cortia, Arisaema* and *Fritillaria*. The aerial portions of many of these plants die off during severe winter, when there is heavy snowfall, leaving the underground parts.

4. Alpine Vegetation

Also termed as 'Alpine scrubs and meadows'. Alpine vegetation in E. Himalaya occurs between 4200-5500 m. The vegetation is strikingly composed of low shrubs and herbs, which are all stunted. *Rhododendron anthopogon* D.Don, *R. campanulatum* D.Don, *R. nivale* Hk. f. (the smallest of the rhododendrons – only 5 cm high), *R. pumilum* Nutt., *Primula sikkimensis* Hk. f., *P. glabra* Klatt, *Meconopsis nepalensis* DC., *Sedum* spp., *Bergenia* spp. *Polygonum* spp., *Pedicularis* spp., *Ephedra* spp. are common. *Saussurea gossypiphora* D.Don is a striking woolly herb. *Thylacospermum caespitosum* (Camb.) Schischk. which forms large hard hemispheric cushions are striking plants in the Arunachal Himalaya. *Leontopodium himalayanum* DC. is also common in this region. The alpine region of Nathu La region of Sikkim is more humid and is characterised by gregarious patches of the ornamental *Primula calderiana* Balf. f. & Cooper and *Rheum nobile* Hk. f. & Th. (the dry leaves of this species are used as a substitute of tobacoo). The alpine zone in Zemu valley of Sikkim contains several ornamental species belonging to the genera *Anemone, Corydalis, Draba, Potentilla, Saxifraga, Sedum, Cortia, Nardostachys, Cremanthodium, Saussurea, Diapensia, Anaphalis, Rhododendron, Cassiope, Primula, Androsace, Gentiana, Swertia, Picrorhiza, Lagotis, Pedicularis, Polygonum, Rheum, Salix, Poa, Juncus* and *Carex.*

The alpine vegetation of south-east Sikkim shows a strong admixture of western Himalaya and Tibetan species, particularly belonging to the families Arecaceae, Primulaceae, Asteraceae, Crassulaceae, Saxifragaceae, Rosaceae and Gentianaceae.

The discussion of the vegetational diversity of eastern Himalaya would not be completed without a reference to the archaic nature of the flora and the nature of its disjunct distribution. Eastern Himalayan flora is regarded as a natural sanctuary of ancient angiosperm taxa. *Magnolia pterocarpa* Roxb. is perhaps the most ancient species of living angiosperm which occur in Arunachal Pradesh. The other primitive taxa of phylogenetic importance are *Magnolia griffithii* Hk.f. & Th. (Assam and Burma) and *Tetracentron sinense* Oliv. (Eastern Himalaya, Burma, China). Eastern Himalaya has a higher endemism than the Western Himalaya due to their isolation because of the lofty mountain ranges and dry Tibetan plateau to the north, warm alluvial plains of the Brahmaputra in the south which act as barriers of plant migration.

Among other interesting elements of the flora worth mentioning are the *Sapria himalayana* Griff., a close relaive of the well known *Rafflesia arnoldii* R. Br. of Malaya (which has the largest flower 15-90 cm across weighing 5 kg.). This species was discovered by Griffith in Mishmi hill in 1836. This curious stemless, leafless parasite is found on the roots of other trees. The flowers are 35 cm across. The species is highly endangered today. *Meconopsis betonicifolia* Franch., a striking blue poppy is also distributed in southeast Tibet, Burma and Yunnan. *Phyllostachys bambusoides* Sieb. and Zucc. is a climbing bamboo of rare occurrence. Among the numerous interesting orchids, *Paphiopedilum fairieanum* (Bl.) Stein. is worth mentioning. Popularly referred to as the "Lost Orchid" this species is recently collected from Kameng. Another orchid *Galeola falconeri* Hk. f. which grows upto 3 m in height is the tallest orchid in India. Besides the above, East Himalaya also abounds in rare, interesting ferns and many gymnosperms with a discontinuous distribution.

For a further discussion on the floristic diversity of the region see under North-East biogeographic zone.

Chapter 6

PHYTO SOCIOLOGICAL REGION OF THE NORTH-EAST INDIA

The north-east Indian phyto sociological region is the most significant one and represents the transition between the Indian, Indo-Malayan, Indo-Chinese phyto sociological regions as well as a meeting place of Himalayan mountains with that of Peninsular India. Therefore, this region acts as a biogeographic gateway for

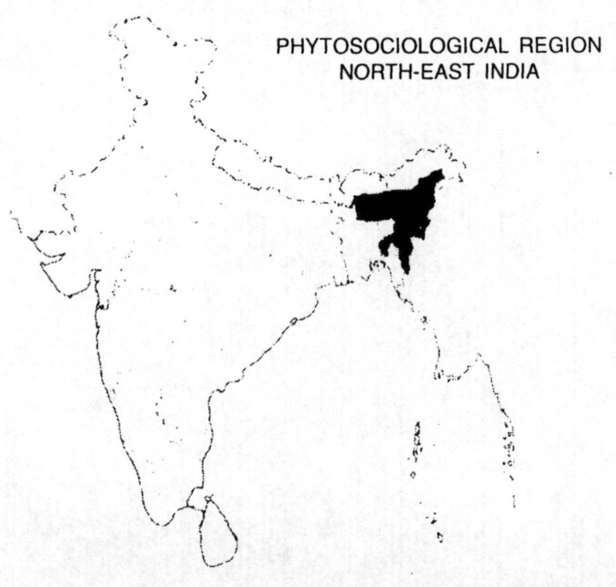

plant migration. This region is the richest in the biological diversity – diversity at the community level, at the species level and in endemics. Most of the species contributing to the biological diversity of north-east India are either restricted to the region as a whole or even to smaller localities as in Khasi and Jaintia Hills, a set of twin hills in Meghalaya which are perhaps the richest botanical habitats in entire Asia.

This, as recognised here, mainly includes all the north-eastern states, excepting Arunachal Pradesh viz. Assam, Manipur, Meghalaya, Mizoram, Nagaland and Tripura, with a total surface area of 171423 sq. km. The region experiences heavy rainfall, frequent flood and landslides. The average annual rainfall in the sub-Himalayan ranges in Assam, Manipur and Tripura is 300 cm, but the rainfall pattern in the region is highly variable, *i.e.* in south Meghalaya at Cherapunji and Mawsynram (the world's rainiest spots) the average annual precipitation of over 11,000 mm has been recorded. But Shillong located only 50 km to the north of Cherrapunji with a rain shadow effect gets only 2300 mm as average rainfall. Much of the rainfall received in the region is through south-west monsoon during June-September. The rainy season is characterised by high humidity conducive for tree growth. There is also a significant variation in the temperature pattern. In general March-June are said to be the summer months. The, advent of summer, is marked by thunder storms and hailstorm accompanied by strong wind. The summer temperature over the hills vary from 5°-30°C and those over the foothills have the range 12°-35°C and in the valley and plains a few degrees more. The winter moths (December-March) are characterised by heavy dew fall and misty nights, while in the higher elevations frost is observed. Thus, the climate within the biogeographic region is highly variable-warm subtropical in the foothills, moderate in the mid slopes, cool and temperate type in the high hills. The varied climate and the altitude have greatly influenced the rich diversity of vegetation in this region.

Vegetational Diversity

The vegetation and flora of north-east India has been studied by several workers, chiefly by Griffith (1847), Clarke (1889), Rowntree (1953), Kanjilal *et. al.*, (1934-40), Bor (1938, 1942), Fischer

(1938), Biswas (1941, 1943), Das (1942), Kingdon Ward (1960), Deb (1960), Naik and Panigrahi (1961), Panigrahi and Joseph (1966), Rao and Joseph (1965), Das and Rajkhowa (1968), Rao (1970, 1974), Haridasan and Rao (1985-87), Rao and Hajra (1986), Jamir and Rao (1988). Based on these works the vegetation of northwest India is classified under 4 types, *i.e.* tropical, subtropical, tremperate and subalpine vegetation, each comprising numerous subtypes, primarily based on altitude and climate factors.

I. Tropical Vegetation

The tropical vegetation in the north-east region typically covers the areas upto elevation of *ca* 900 m. Depending upon the locality and the degree of human interference, several kinds of forests like (1) evergreen and semi-evergreen forests, (2) deciduous forests, (3) grasslands, (4) riparian forests and swamps are seen.

(i) Tropical evergreen and semi-evergreen forests

Characteristically this receives very high rainfall and high humidity. These forests represent one of the major ecological types in the zone with a rich floristic diversity. The vegetation seems to be of a climatic climax type.

There is a bewildering wealth of species in these multitiered forests. The top canopy normally comprises some of the lofty tree species such as *Dipterocarpus retusus* Bl., *D. turbinatus* Geartn., *Terminalia myriocarpa* Heurck & Muell., *Duabanga grandiflora* (Roxb. ex DC.) Walp., *Artocarpus chama* Buch.-Ham., *Tetrameles nudiflora* R. Br., *Ailanthus integrifolia* ssp. *calycina* (nees) Nees, *Altingia excelsa* Noronha, *Kayea assamica* (Pierre) Nooteb., *Mesua ferrea* L., *Phobe lanceolata* King & Prain, *Aglaia hiernii* Vishwa. & Ramach., *Dysoxylum binectariferum* Hk.f., *Gmelina arborea* L.

The middle storey is characterised by the preponderance of *Mesua ferrea* L., *Endospermum diadenum* (Miq.) Airy Shaw, *Garuga pinnata* Roxb., *Ficus rumphii* Bl., *Lagerstroemia parviflora* Roxb., *Syzygium grande* Walp., *Gardenia paniculata* Roxb., *G. pedunculata* Roxb., *Ficus racemosa* L., *Sapium baccatum* Roxb., *Sterculia hamiltonii* (Kuntze) Adelb. In certain semi-evergreen patches this layer is however dominated by *Bischofia javania* Bl., *Dysoxylum gobora* (Bunch.-Ham.) Merr., *Vatica lanciaefolia* Bl., and *Cyclostemon elliptium* Hk.f.

The thrid storey wherever evident, consists of trees of 5 to 10 m high, and the main components in this category are *Premna bengalensis* Cl., *Carallia brachiata* (Lour.) Merr., *Grewia disperma* Rottl., *Hibiscus macrophyllus* Roxb., *Picrasma javanica* Bl., *Ficus racemosa* L., and a number of smaller trees of the higher storey. The shrub layer is quite evident and often merges with smaller tree layer. The common ones are *Leea asiatica* (L.) Ridsd., *Abaroma angusta* L., *Boehmeria macrophylla* D.Don, *Ficus hispida* L., *Dracaena spicata* Roxb., *Ardisia paniculata* Roxb. and a few others. The herbaceous flora is also very profuse mainly belonging to Rubiaceae, Zingiberaceae, Begonicaceae, Fabaceae, Poaceae and Acanthaceae. Among the Zingiberaceae memebrs *Curcuma* spp., *Curcumorpha longiflora* (Wall.) Rao & Verma, *Phrynium capitatum* Willd., *Costus speciosus* Sm. and *Zingiber zerumbet* Sm. are the important ones. In shady places there are associations of *Amischotolype molissima* (Bl.) Hassak., *Colocasia esculenta* (L.) Schott, etc. In marshy places and along the streams *Angiopteris* sp. characteristically appears. Many of the species of the ground cover are quite attractive on account of their showy flowers or foliage. Tree ferns (*Cyathea*) and handsome *Angiopteris* are remarkable. Lianas (woody climbers) like *Ampelocissus latifolia* (Roxb.) Planch., *Cayratia pedata* (Lour.) Gagnep., *Entada phaseoloides* (L.) Merr., *Bauhinia vahlii* Wt. & Arn., *Derris trifoliata* Lour., *Combretum latifolium* Bl., *Fissistigma* sp., *Mezoneurum* sp. and *Gnetum montanum* Mfg. are common. Several species of the prickly *Calamus* stretching for long distances over the tree canopy, particularly in moist area is also a common feature. In moist areas and in valleys, oftern these forests are interspersed by bamboo species of *Dendrocalmus hamiltonii* Nees & Arn., and *Bambusa tulda* Roxb.

The epiphytic species also show a great profusion and variety in these forests. Of these, broadly the ferns and orchids are quite conspicuous. Among the orchids *Dendrobium* spp. and *Cymbidium* spp. top the list. The other epiphytic species are *Aeschynanthus acuminata* Wall., *Peperomia pellucida* (L.) H.B.K. and species of *Agapetes, Hoya* and *Medinella.* In certain places, particularly in the subtropical zone there are very many Loranthaceous parasites belonging to *Helixanthera, Taxillus, Viscum* and *Dendrophthoe.* In extremely moist habitats and in places where water is dripping, insectivorous plants like *Utricularia bifida* L., *U. striatula* Sm. etc.

also come up. The epiphytic climbers like *Pothos, Rhaphidophora* sp. and *Scindapsus officinalis* Schott and the interesting *Dischidia rafflesiana* Wall. (very rare) also belong to this group. Among the ferns *Pseudodrynaria coronans* (Wall.) Ching, *Microsorum punctatum* (L.) Copel, *Pyrrosia* spp., *Lepisorus* spp., *Polypodium* spp. and *Lycopodium* spp. are important. In certain places moss covered tree trunks are so profusely covered by epiphytes that they almost look like haging gardens.

The diversity and species composition vary from place to place within the same region. Yet in other places some of the species listed above are totally absent. Another point to be noted is that at some places even the deciduous elements enter the area, perhaps due to biotic interference.

(ii) Tropical deciduous forests

Tropical deciduous forests occur in areas with less than 150 cm rainfall. Most of the deciduous forests in the northeast are not the typical natural deciduous forests but are only subclimax, man - made forests. These forests are characterised by seasonal leaf shedding and profuse flowering of trees. Forest fires are also common. Deciduous forests in north-east India are quite extensive and are a home of many valuable timber trees, such as *Shorea robusta* Gaertn. f. (sal), *Tectona grandis* L.f. (teak). The deciduous forests of north-east India are predominated by *Shorea robusta* Gaertn.f. which are also commercially exploited. The other important species of the forests are *Tectona grandis* L.f., *Sterculia villosa* Roxb. ex Sm. and *Gmelina arborea* L. These species come up as natural as well as planted. *Schima wallichii* (DC.) Korth, *Tetrameles nudiflora* R. Br., *Lannea coromandelica* (Houtt.) Merr., *Bombax ceiba* L., *Vitex peduncularis* Wall ex Sch., *Albizia* spp. are also in abundance but in scattered patches. These trees of the deciduous canopy are always lofty and straight boled with spreading crowns.

A distinct second storey composed of *Aporusa octandra* (Buch.-Ham. ex D. Don) Vickery, *Careya arborea* Roxb., *Rhus semialata* Murr., *Bridelia retusa* Baill., *Grewia microcos* L., *G. disperma* Rottl., *Mallotus ferrugineus* (Roxb.) Muell.-Arg., *Glochidion* spp. are also not uncommon. These trees show a narrow crown and are laxly branched.

The shrubby layer is often gregarious and form impenetrable thickets during rainy season with profuse growth of spreading shrubs interwoven by annual climbers. The main components of this layer are *Desmodium pulchellum* Benth., *Flemingia macrophylla* (Willd.) Prain, *Holarrena antidysenterica* Wall. ex A. DC., *Leea asiatica* (L.) Ridsd., *Glycosmis arborea* DC., and *Licuala peltata* Roxb.

The lianas are fewer. These belong to *Combretum* spp., *Hiptage benghalensis* (L.) Kurz., *Spatholobus parviflorus* (Roxb.), Kuntze, *Bauhinia nervosa* (Benth.) Baker, *Entada pursaetha* DC. They make their way high up the canopy and become conspicuous, but often deforming the host trees.

The epiphytic flora is extremely low as compared to the evergreen forests. However, massive growth of certain orchids like *Pholidota imbricata* Lindl., *Dendrobium* spp., and *Papilionanthe teres* (Roxb.) Schltr. and at certain places *Rhynchostylis retusa* (L.) Bl. densely cover the tree trunks with festoons of flowers. The tree trunks also provide a favourable habitat for the growth of several epiphytic ferns, like *Drynaria* spp., *Microsorum* spp. and the biologically interesting *Dischidia nummularia* R.Br. of the family Asclepiadaceae.

The herbaceous undergrowth in these forests is also quite variable. Mainly the introduced weedy species of the genera *Eupatorium, Mikania, Galinsoga* and *Lantana camara L.* can be seen during the dry period when the other annual species dry up. Added to this, the recurring forest fires further add to the elimination of the herbaceous flora.

In some places, the deciduous forests tend to become mixed deciduous forests with the intrusion of evergreen elements such as *Elaeocarpus floribundus* Bl., *Toona ciliata* Roem., *Castanopsis tribuloides* (Sm.) DC., etc. A good number of bamboos also grow interspersed amidst these forests particularly in open places.

Maximum human interference is noticed in these forests. Where such dense natural forests are degraded, trees like *Bauhinia purpurea* L., *Macaranga denticulata* Muell., *Callicarpa arborea* Roxb., *Emblica officinalis* Gaertn., *Careya arborea* Roxb., *Grewia microcos* L., *Holarrhena antidysenterica* Wall. ex A.DC., *Ziziphus oenoplea* Mill. are noticed.

(iii) Tropical Bamboo forests

Tropical bamboo forests are not natural but appear in Jhum fallows of 25-50 years. These forests at places form pure stands. The common species are *Dendrocalamus hamiltonii* Nees ex Arn., *D. giganteus* Munro, *Drepanostachyum khasianum* (Munro) R. Majumdar, *Melocanna baccifera* (Roxb.) Kurz, *Schizostachyum dullooa* (Gamble) R. Majumdar and *Bambusa tulda* Roxb. While the former 3 species prefer moist localities along streams, *Bambusa tulda* Roxb. prefers well drained hill slopes. Although about 100 species of bamboos are reported from north-east India many of them have become quite rare due to the unscrupulous exploitation for the immediate economic gains.

(iv) Tropical Grasslands/Savannahs

Tropical grasslands or Savannahs occur in riparian flats *inundated* by flood water of the river Brahmaputra. The grasses are tall and belong to the species of *Saccharum, Anthisteria, Erianthus, Arundo donax* L., *Phragmites austrialis* (Cav.) Trin. ex Steud., *Panicum atrosanguinium* A. Rich., *Setaria* spp., *Arundinella bengalensis* (Spreng.) Druce and a few others. These grasslands are also not climax types but are as a result of removal of original forest cover. The tropical grasslands are distinct from those of higher altitudes of the Shillong plateau, where the rolling grasslands are composed of much shorter grasses at the ground level. In many places these grasses are intermixed with sedges.

Apart from giving a green look to the barren hills, these rolling grasslands also support some of the other dicotyledonous species, many of which are rainy season weeds. Some the these are *Eriosema himalaicum* Ohashi *Polygonum bistorta* Garcke, *Trifolium repens L.*, *Sonchus asper* Will., *Eusteralis linearis* (Benth.) Panig., *Osbeckia stellata* D. Don, *Hypochaeris radicatà* L., *Hemiphragma heterophyllum* Wall. and *Drosera peltata* Sm.

The Grassland area of the Kaziranga National Park represents a combination of grasslands, swamp forests and marsh, which form the ideal habitat for the one horned rhino. Tall grasses reaching *ca* 5 m such as *Erianthus longisetosus Andr., Arundo donax* L., *Thysanolaena maxima* Kuntze *Saccharum spontaneum* L., *Vetiveria zizanioides* Nash and *Phragmites austrialis* (Cav.) Thrin. ex Stud. are common. Occasionally, these tropical grasslands support scattered tree species such as *Semecarpus anacardium* L.f., *Albizia* spp., *Butea monosperma* (Lam.) Taub., *Lagerstroemia speciosa* (L.) Pers. and *Duabanga grandiflora* (Roxb. ex DC.) Walp.

(v) Swamp or march vegetation

A singular feature of the tropical vegetation in the warm humid Assam valley is the swamp or marsh vegetation (Rao, 1974). The Brahmaputra valley still contains extensive patches of natural vegetation in the form of swamps and marshes. There are innumerable stagnant ponds in the region, specially in Golpara, Kamrup and North Lakhimpur. These ponds are locally called 'Beel' support a rich diversity of aquatic angiosperms mainly belonging to the families Nymphaeaceae, Lemnaceae, Araceae, Cyperaceae, Poaceae, Eriocaulaceae and Najadaceae. *Euryale ferox* Salisb. with large orbicular leaves covering the water surface of the ponds is very common in Kamrup district. The other species are *Typha elephantina* Roxb., *Arundo donax* L., *Phragmites australis* (Cav.) Trin. ex Steud., shrubs of *Crateva magna* (Lour.) DC., *Syzygium cuneatum* (Duthie) Balakr., *Homonoia riparia* Lour. form the characterisitc vegetation along the river channels and banks of ponds etc.

II. Subtropical Forests

The subtropical forest typically covers the elevaton from 900-1800 m with an average annual rainfall of 150-500 cm. The forest types are mainly–

(i) Subtropical evergreen and semi-evergreen forests

These forests resemble in many respects to those of tropical types and in certain places there is even overlapping of species. These forests are also climatic climax forests seen scattered in valleys, banks of rivers and strems and in pockets on hills. The trees are generally of a bushy appearance and shorter than in the tropical zone. The stratification is also not clear. The shrubby and herbaceous layers are well marked. The lianas are comparatively fewer in number, but the apiphytes are many. The undergrowth is almost impenetrable. The diversity of the species is quite enormous. Some of the common tree species are *Castanopsis tribuloides* DC.,

Engelhardtia spicata Bl., *Ficus elastica* Roxb. ex Hornem., *Magnolia insignis* (Wall.) Bl., *Exbucklandia populnea* (R. Br. & Griff.) R.Br., *Elaeocarpus floribundus* Bl., *Prunus napaulensis* Koch, *Radermachera gigantea* Miq. and *Ehretia acuminata* R. Br. These trees at certain places merge with trees of second storey composed of *Micromelum minutum* (Forst. f.) Wt. & Arn., *Gardenia latifolia* Ait., *Symplocos paniculata* (Thunb.) Miq., *Michelia punduana* Hk. f. & Th., *Ternstroemia gymnanthera* (Wt. & Arn.) Bedd. and several others. Tree ferns are also common in these forests. Herbaceous and shrubby layers are very dense. The common shrubby speceis belong to the species of *Sarcococca, Neillia, Ixora, Eurya, Psychotria, Ardisia* and *Camellia.* Among the herbaceous flora apart from the fungi, mosses, *Selaginella* the and *Lycopodium,* various angiosperm species belonging to *Begonia palmata* D. Don, *Houttuynia cordata* Thunb., *Sonerila* sp., *Impatiens* spp., *disporum* spp. and various speceis of the family Zingiberaceae, Araceae, Commelinaceae and Gesneriaceae. Terrestrial orchids and root parasites of *Balanophora* spp. are also common.

The lianas are fewer in number compared to the evergreen zone. *Mucuna macrocarpa* Wall., *Tetrastigma odoratum* (Laws.) Gagnep., *Celastrus* spp., etc. can be seen.

(ii) Subtropical pine forests

Subtropical pine forests are found at higher elevation (1200-2000 m) along the slopes of hills. These are climax foress of secondary nature. The average annual rainfall is around 175 cm. *Pinus kesiya* Royle ex Gourd. (Khasi Pine) is the principal element occurring in almost pure stands. Occasionally broad leaved species such as *Schima wallichii* Choisy, *Rhododendron arboreum* Sm., *Ternstroemia gymnanthera* (Wt. & Arn.) Bedd., *Engelhardtia spicata* Leschen. ex Bl. and species of the family Lauraceae are also associated. The forest floor, covered by a thick mat of pine needles also supports the growth of shrubby and herbaceous species. Some such species are *Lyonia ovalifolia* (Wall.) Drude, *Meizotropis buteiformis* Voigt, *Dipsacus asper* DC., *Inula cappa* D.C., *Saussurea crispa* Vaniot, *Anaphalis contorta* Hk. f., *Osbeckia* spp., *Vernonia saligna* DC., *Rubus rugosus* Sm., *Artemisia japonica* Thunb. and *Callicarpa rubella* Lindl. Among the herbaceous ones *Scutellaria discolor* Coleb., *Sanicula elata* Ham., *ainsliaea latifolia* (D. Don) Sch., *Ajuga macrosperma* Wall. ex Benth., *Neanotis wightiana* (Wt. & Arn.) Lewis, *Valeriana jatamansi* Wall., *Hedyotis corymbosa* (L.) Lam., *Selinium striatum* Benth., *Senecio griffithii* Hk. f. & Th. and the curious *Aeginetia indica* L. a parasite on the roots of grasses, are quite frequent in the pine forests.

III. Temperate Forests

The temperate vegetation chiefly occupies the areas between 1800-3500 m and are commonly seen in Shillong plateau, Nagaland, Mizo and Mikir hills. The rainfall here is very high (200-500 cm) with a severe winter during November-March. The ground forest is also common during December-January.

These climatic climax forests are usually found in isolated pockets along valley slopes and strips. In comparatively lower region, mixed forests of *Acer, Betula, Juglans, Magnolia, Michelia, Quercus, Rhododendron* and others characterise the hill tops and the valleys. At higher elevations the temperate forests are dominated by the genera *Rhododendron, Pinus* and *Tsuga*. The temperate conifer forests found mostly in Naga hills and Manipur are chiefly of *Pinus wallichiana* Jack. associated with *Rhododendron, Quercus, Lyonia, Ilex* sp., *Alnus nepalensis* D. Don, *Cornus controversa* Hemsl ex Prain, and a few other broad leaved species. In Meghalaya and adjoining regions the temperate vegetation generally show a bushy and stunted habit. They form a dense canopy with main components like *Lithocarpus fenestrata* (Roxb.) Rehd., *Panax pseudo-ginseng* Wall. and members of Commelinaceae, Araceae and the saprophytic *Cheilotheca humilis* (D. Don) Keng (*Monotropa uniflora* L.) a unique component of this forest. The epiphytic flora is exceptionally rich as almost all the trees are heavily clothed with a layer of epiphytes. The fern flora dominates over flowering plants. These belong to *Asplenium nidus* L., *A. normale* D. Don, *Loxogramme involuta* (D. Don) Presl., *Polypodium argutum* Wall. ex Hk., *Lepisorus* spp., *Pyrrosia* spp. and several fern allies such as *Lycopodium* spp. Among the epiphytic orchids mention may be made of *Coelogyne punctulata* Lind., *Pleione praecox* (Sm.) D. Don and *Dendrobium* spp., *Agapetes obovata* Hk.f., *Vaccinium sprengelii* (G. Don) Sleumer, *Aeschynanthes* spp., *Castanopsis kurzii* (Hance) Biswas, *C. ornata* Spach., *Quercus griffithii* Hk.f. & Th. *Myrica esculenta* Buch.-Ham. ex D. Don, *Symplocos glomerata* King, *Photinia arguta* Wall. ex Decne., *Sloanea dasycarpus* (Benth.) Hemsl., *Elaeocarpus* spp., *Magnolia insignis* (Wall.)

Bl., *Ilex* spp. and a host of others form an unparallel assemblage in varied proportion. A well developed shrub layer can also be seen and this layer is usually composed of *Mahonia pycnophylla* (Fedde) Takeda, *Daphne papyracea* Wall., *D. involucrata* Wall., *Viburnum* spp., *Camellia* spp. and *Lyonia ovalifolia* (Wall.) Drude. The forest floor has often a gregarious growth of ferns particularly of *Plagiogyria* spp. and *Dryopteris* spp.

Climbers are usually few and less frequent. Some of these are straggling shrubs belonging to *Aspidopterys indica* (Willd.) Hochr., *Rourea minor* (Gaertn.) Alston, *Rosa brunonii* Lindl., *Kadsura heteroclita* (Roxb.) Craib., *Elaeagnus pyriformis* Hk.f., *Ficus laevis* Bl., *Hedera nepalensis* Koch, *Codonopsis javanica* (Bl.) Hk.f., *Porana racemosa* Dalz. & Gibs.

Sacred forests

The temperate vegetation in the Khasi and Jaintia hills in the form of "Sacred forests" at Shillong peak, Mawphlong and Mowsmai deserve a special mention in the discussion of the temperate forests of north-east India. These forests are the relict types and are left in small pockets untouched due to religious beliefs and myths (Bor, 1942). These pockets of dense forests in saucer-shaped depressions offer us a glimpse of the original forests that must have once clothed these hills. They are a rich store house of vegetal wealth uncomparable to any other type of forest in the region. Many of the rare and endangered species of the region are now finding a refuge in these sacred forests. Some of the dominant species in these sacred forests belong to the genera *Castanopsis, Photinia, Eriobotrya, Pyrus, Prunus* and *Sorbus*. The primitive flowering plants like *Corylopsis himalayana* Griff. and *Exbucklandia populnea* (R.Br. ex Griff.) R.Br., *Mangleitia* sp. and the climbing *Schizandra* and *Kadsura* are also frequent. Apart from numerous epiphytic orchids and ferns, the family Vacciniaceae with *Agapetes* and *Vaccinium* are very common on almost all trees. The forest floor has a thick mat of litter. In contrast to the surrounding disturbed vegetation, these protected forests offer us altogether a different picture.

IV. Subalpine Vegetation

The subalpine vegetation in north-east India have a very limited distribution and are quite different from those which are seen in the Himalayan region. The subalpine vegetation mainly occurs at 3500-4200 m altitude in the Naga hills and in Manipur. The dominant tree species are *Abies spectabilis* Spach, *Picea spinulosa* (Griff.) Henry and *Tsuga* sp. Shrubby stunted species of *Rhododendron, Juniperus, Berberis, Cotoneaster, Salix* and others are also common. The herbaceous species are also in profusion and exhibit enormous diversity. These mainly belong to the genera of *Aconitum, Anemone, Pedicularis, Potentilla* and *Primula*.

Plant Diversity

The north-east India including Eastern Himalaya are positioned in such a way that the region is not only able to capture maximum precipitation and high humidity, which is conducive for flora, but also comes in direct contact with many other floristic regions, so that there is a free migration of flora. We find in this region a representation of tropical, temperate, subalpine and alpine vegetation, with their characteristic species composition. All these factors have rendered this region as the richest botanical diversity centre in the entire subcontinent.

The following facts should highlight the floristic diversity of the north-east India (including east Himalaya). The total number of the flowering plant species in India is expected around 16,000 of which about 50 per cent (8000 species) of the species hail from this region. Out of about 315 flowering plant families in India more than 200 families are represented in this region. The Table 6.1 shows the number of genera and species (approximate) occurring in India and north-east India with respect to some dominant families of flowering plants.

Many families, represented in India by a solitary genus with 1 or 2 species are represented in this region, *e.g.* Coriariaceae, Nepenthaceae, Turneraceae, Illiciaceae, Ruppiaceae, Siphonodontaceae, Tetracentraceae, and a few others (Rao and Murti, 1990). Similarly a number of genera in India with 1 or 2 species are also represented here.

Out of 3 genera and 36 species of Elaeocarpaceae in India 2 genera and 23 species are found in this region. The family Elaeagnaceae with 15 species in India is represented by 13 species in E. Himalaya. There are other interesting features of the flora

Table 6.1 : Approximate number of genera and species in some dominant families of higher plants

Name of family	N.E. Region	India	World
1. Poaceae	160/500	240/1100	620/10000
2. Orchidaceae	104/700	200/1500	735/20000
3. Fabaceae	50/200	100/750	482/12000
4. Caesalpiniaceae	11/42	23/801	52/2800
5. Mimosaceae	10/35	15/75	56/2800
6. Asteraceae	25/70	135/710	900/13000
7. Cyperaceae	14/175	21/350	90/4000
8. Lamiaceae	30/95	65/380	180/3500
9. Scrophulariaceae	15/35	60/350	220/3000
10. Acanthaceae	25/125	70/340	250/2500
11. Euphorbiaceae	55/160	65/340	300/5000
12. Rubiaceae	50/170	80/280	500/6000
13. Urticaceae	15/45	25/114	45/550
14. Zingiberaceae	18/73	20/115	46/850

Table 6.2 : The number of species in some selected taxa/groups in India and in north-east India.

Genera/group	*India*	*North-east India*
Dendrobium	75	65
Bulbophyllum	50	35
Liparis	45	35
Ceologyne	35	30
Habenaria	100	45
Paphiopedilum	5	4
Vitis	70	46
Citrus	11	10
Bamboo	100	58
Rhododendron	90	80
Hedychium	40	34
Elaeocarpus	30	23
Echinocarpus	5	4
Elaeagnus	12	10
Hippophae	3	3
Ferns & Fern - allies	1000	600
Haplophragma	1	1
Harpullia	1	1
Nepenthes	1	1
Circaester	1	1
Haematocarpus	1	1
Haloragis	2	1
Helwingia	2	1

contd...

Table 6.2 contd...

Genera/group	*India*	*North-east India*
Hemiboea	1	1
Hemsleya	1	1
Holboellia	1	1
Hypecoum	1	1
Hymenopogon	2	1
Hymenocardia	1	1
Hymenandra	1	1
Hymenachne	2	1
Achyrospermum	1	1
Acanthopanax	2	1
Ichnanthus	1	1
Illicium	1	1
Iodes	2	1
Keenania	1	1
Nyssa	2	2
Wikstroemia	2	2
Willughbeia	1	1

worth mentioning. *Cycas pectinata* Griff. a rare species occurs scattered in Kamrup. *Gnetum gnemon* L. has been recorded from the Khasi, Jaintia and Naga hills. *Podocarpus nertifoliius* D.Don also occurs in the Khasi hills. *Euryale ferox* Salisb. with giant floating leaves has its distribution in several districts of Assam.

North-east India shares the maximum number of endemic species and other rare plants showing discontinuous distribution. Some such endemic species are *Uvaria lurida* Hk.f. & Th., *Magnolia gustavi* King, *M. griffithii* Hk. & Th., *Pachylarnax pleiocarpa* Dandy, *Ilex embellioidea* Hk. f., *Distylium indicum* Benth., *Merrilliopanax cordifolia* Sastry, *Ardisia quinquangularis* A. DC., *A. rhynchophylla* Cl., *Micholitzia obcordata* N.E. Br., *Acanthophippium sylhetensis* Lindl., *Aphyllorchis vaginata* Hk.f., *Eria barbata* (Lindl.) Reichb. f., *Gastrodia exilis* Hk. f., *Paphiopedilum insigne* (Wall.) Pfitz., *Hedychium calcaratum* Rao & Verma, *H. dekianum* Rao & Verma, *H. marginatum* Cl. and *Nepenthes khasiana* Hk.f. *Nymphaea pygmaea* Ait. (Siberia, N. China), *Michelia velutina* DC. (Nepal, Tibet, Burma), *[illegible]ium cambodiana* Hance (Southern Indo-China), *Homalium schleichii* Kurz (Burma), *Salomonia aphylla* Griff. (Burma), *Anneslea fragrans* Wall. (Burma), *Cotylanthera tenuis* Bl. (Java), *Mitrastemon yam[illegible]ctoi* Makino (Japan, Sumatra), *Aphyllorchis montana* Reichb. f. (Sri Lanka), *Dendrobium bensoniae* Reichb. f. (Burma, Thailand), *Epipogium roseum*

(D. Don) Lindl. (W. Africa, Java, Australia), *Polystachya flavescens* (Bl.) Sm. (Java, Africa, S. India), etc. are some taxa showing discontinuous distribution.

Nepenthaceae, with the single endemic species, *Nepenthes khasiana* Hk.f. represents the northernmost limit of this family, with a general range of distribution from Madagascar to Malaysia. Similarly *Zeylanidium* of Podostemaceae, found in the Kameng District of Arunachal Pradesh, also represents the northernmost distribution of the family in India. The occurrence of *Alniphyllum fortunei* (Hemsl.) Makino and *Huodendron biaristatum* (Sm.) Rehd. in Subansiri district of Arunachal Pradesh is of considerable significance. The known distribution of the former is Yunnan and China and of the latter is Burma and Tonkin. *Munronia pinnata* Wt. occurs in Sikkim and Khasi Hills and also in the Western Ghats of South India. *Coptis teeta* Wall. is endemic in Arunachal Pradesh. The primitive family Magnoliaceae has most of its members only in the eastern Himalaya and Assam. The related family Annonaceae is mostly limited to north-eastern region and the Deccan Peninsula.

Certainly the floristic diversity in the East Himalaya is enormous. It is not possible to highlight the diversity in all the groups and as such some selected taxa/groups are discussed here.

Orchid diversity

Orchids which are well known for their showy and long lasting flowers exhibit a remarkable diversity in the east Himalaya. The Orchidaceae is the second largest family of flowering plants in India. Pradhan (1976) estimated *ca* 125 genera and 800 species in India whereas Jain and Mehrotra (1984) list 144 genera and 925 species. Considering the recent additions to the Orchid flora one can conveniently put the number of orchids at 1000, of which *ca* 700 species are represented in the east Himalaya alone. Some of the larger genera having maximum diversity are listed in Table 6.2. The humid forests of the east Himalaya provide a very suitable habitat for the growth of both terrestrial and epiphytic orchids. There is also such a great profusion of epiphytic orchids in certain forests, that having covered completely the host trees, the orchids fall down and establish on the litter itself. The author during one of his field trips in Khasi hills counted 18 species of orchids just on one tree itself. With the removal of such forest trees the orchids are also disappearing in the same magnitude. As many as 300 species of orchids have already become endangered in the E. Himalaya.

Rhododendron diversity

The genus *Rhododendron* of Ericaceae is another remarkable group of showy plants which show maximum diversity in the Himalaya. The genus has about 90 species in India (Himalaya) of which 80 species are exclusively confined to east Himalaya. Arunachal Pradesh alone has *ca* 70 species. Sikkim is another region rich in *Rhododendron*. Recently a *Rhododendron* sanctuary has also been identified here. Apart from just the number of species, the genus has life form diversity like herbs, shrubs and trees. *Rhododendron* nivale Hk.f. is the smallest of the rhododendrons in India, which measures only 5 cm. Some of the common rhododendrons are *R. anthopogon* D. Don, *R. arboreum* Sm., *R. campanulatum* D. Don, *R. hodgsonii* Hk.f., and *R. thomsonii* Hk.f.

Diversity in the genus Hedychium

The genus *Hedychium* of the family Zingiberaceae is another group of ornamental plants that can be directly introduced into our gardens. The fragrant flower on terminal spikes have attractive white, yellow, orange, and red colours. There are *ca* 40 species in India, of which 35 species occur in east Himalaya alone. From Arunachal Pradesh itself 18 species are reported (Haridasan and Hegde, 1991). Some of the common species of the genus are *H. aurantiacum* Rosc., *H. coronarium* Koen., *H. densiflorum* Wall., *H. thyrsiforme* Ham. ex Sm. and *H. villosum* Wall; the rare ones are *H. luteum* Angl. ex Link, *H. greenii* W.W. Sm. *H. aureum* Cl., *H. longipedunculatum* Sastry & Verma, *H. radiatum* Rao & Hajra, *H. dekianum* Rao & Verma and *H. Wardii* Fisch.

Diversity in Bamboos

Bamboos play an important role in the economy of the country and are associated with the human kind since ancient times. Tropical Asia including India is the main centre of bamboo diversity. Bamboo forests come up both in tropical and temperate regions and today approximately 12.8 per cent of the total forest area in India is covered by bamboo forests. Out of 18 genera and 130 species so far known in India 15 genera and 63 species are represented. in E. Himalaya as shown below. *Arundinaria* (10 spp.) *Bambusa* (22 spp.) *Cephalostachyum* (7 spp.), *Chimanobambusa* (9 spp.), *Dendrocalamus* (15 spp.), *Thalamnocalamus* (4 spp.), *Dinochloa* (2 spp.) *Gigantochloa* (2 spp), *Melocalamus* (1 sp.), *Malacanna* (2 spp.) *Neohouzeaua* (2 spp.) *Phyllostachys* (2 spp.), Pseudostachyum (1 sp.), *Teinostachyum* (1 sp.), *Oxtenanthera* (2 spp.) (Biswas, 1988). East Himalaya forms the centre of genetic diversity for the genera *Bambusa, Dendrocalamus, Arundnaria* and *Cephalostachyum.* A few important bamboo species are *Arundinaria maling* Gamble, *A. racemosa* Munro, *Bambusa arundinacea* Retz., *B. glaucescens* (Willd.) Seib. ex Munro, *B. mastersii* Munro, B. *pallida* Munro, *B. tulda* Roxb. *B. vulgaris* Schrad, *Cephalostachyum capitatum* Munro, *C. latifolium* Munro, *C. pallidum* Munro, *Chimonobambusa callosa* Nakai, *C. griffithiana* (Munro) Nakai, *C. hookeriana* Nakai, *C. intermedia* Nakai, *Dendrocalamus giganteus* Munro, *D. hamiltonii* Nees & Arn. ex Munro, *D. strictus* (Roxb.) Nees, *Melocanna baccifera* (Roxb.) Kurz, *M. bambusoides* Trin., *Oxytenanthera albociliata* Munro, *Phyllostachys assamica* Gamble ex Brandis, *Teinostachyum griffithii* Munro, *Thamnocalamus aristatus* Camus, *T. spathiflorus* Munro, and *T. falconeri* Hk. f. ex Munro. Commercially *Bambusa arundinacea* Retz., *B. balcooa* Roxb., *B. nutans* Munro, *B. tulda* Roxb., *Dendrocalamus hamiltonii* Nees & Arn. ex Munro, *Thamnocalamus falconeri* Hk. f. ex Munro, *T. spathiflorus* Munro are important.

The genetic resource is getting eroded in the following endangered bamboo species as their populations have become highly fragmented mainly due to the practice of shifting agriculture and selective removal. *Arundinaria clarkei* Hk. f., *A. mannii* Gamble, *Bambusa mastersii* Munro, *Cephalostachyum capitatum* Munro var. *decomposita* Gamble, *Dendrocalamus sahnii* Naithani & Bahadur, *Dinochloa maclellandii* (Munro) Kurz, *Gigantochloa takserah* Camus, *Phyllostachys assamica* Gamble ex Brandis, *Phyllostachys mannii* Gamble, *Pleioblatus simonii* (Carr.) Nakai, *Semiarundinaria pantligii* Gamble, *Sinobambusa elegans* (Kurz) Nakai.

North-east Indian region is also rich in a number of commercial timbers as shown in Table 6.3.

Table 6.3 : Important timber species of North-east region of India

Sl. No.	Species	Trade Name	Family
1.	*Albizia odoratissima*	Black siris	Mimosaceae
2.	*Terminalia bellirica*	Bahera	Combretaceae
3.	*Lagerstroemia microcarpa*	Benteak	Lythraceae
4.	*Phoebe* sp.	Bonsum	Lauraceae
5.	*Bischofia javanica*	Bishop wood	Euphorbiaceae
6.	*Michelia champaca*	Champa	Magnoliaceae
7.	*Schima wallichii*	Chillnni	Theaceae
8.	*Cinnamomum* spp.	Cinnamon	Lauraceae
9.	*Artocarpus chama*	chaplash	Moraceae
10.	*Chukrassia velutina*	Chickrassy	Meliaceae
11.	*Dillenia* spp.	Dillenia	Dilleniaceae
12.	*Gmelina arborea*	Gamari	Verbenaceae
13.	*Ailanthus integrafolia*	Gokul	Simaroubaceae
14.	*Terminalia chebula*	Harra	Combretaceae
15.	*T. myriocarpa*	Hollock	Combretaceae
16.	*Haldina cordifolia*	Haldu	Rubiaceae
17.	*Dipterocarpus maerocarpus*	Hollong	Dipterocarpaceae
18.	*Hopea* glabra	Hopea	Dipterocarpaceae
19.	*Syzygium* spp.	Jamun	Myrtaceae
20.	*Castanopsis pupureilla*	Indian chestnut	Fagaceae
21.	*Anthocephalus chinensis*	Kadam	Rubiaceae
22.	*Careya arborea*	Kumbi	Lecythidaceae
23.	*Mesua ferrea*	Mesua, Nagkesar	Clusiaceae
24.	*Carallia brachiata*	Manigewa	Rhizophoraceae
25.	*Mangifera indica*	Mango	Anacardiaceae
26.	*Shorea robusta*	Sal	Dipterocarpaceae
27.	*S. assamica*	Makai	Dipterocarpaceae
28.	*Persea macrantha*	Machilus	Lauraceae
29.	*Azadirachta indica*	Neem	Meliaceae
30.	*Aphanamixis polystachya*	Pitraj	Meliaceae
31.	*Stereospermum dereonatum*	Padri wood	Bignoniaceae
32.	*Lagerstroemia hypoleuca*	Pyinura	Lythraceae
33.	*Dalbergia sissoo*	Sheesham	Papilionaceae
34.	*Toona ciliata*	Toon	Meliaceae
35.	*Dysoxylum malabaricum*	White cedar	Meliaceae
36.	*Albizia procera*	White siris	Mimosaceae
37.	*Juglans regia*	Walnut	Juglandceae

Medicinal Plant Diversity

The Himalaya in general, since ages have served as a storehouse of humerous life saving drug plants. The numerous adivasi populations inhabiting the densely wooded regions in the east Himalaya have developed an intimate knowledge of medicinal plants after centuries of trial and error method. Recent ethnobotanical studies of the author (Rao, 1981; Rao and Jamir, 1982; Rao and Haridasan, 1991) have brought to light numerous such medicinal plants, with an enormous diversity in them. Only a few important ones are listed here. *Coptis teeta* Wall. (Mishmitita), *Paederia foetida* L. (Gandhali), *Podophyllum hexandrum* Royle (Papra), *Nardostachys grandiflora* DC. (Jatamansi), *Panax pseudo-ginseng* Wall. (Ginseng), *Picrorhiza kurrooa* Royle ex Benth. (Katki), *Hydnocarpus kurzii* (King) Warb. (Chaulmugra), *Alpinia galanga* Wild. (Bara kulapjan), *Elaeagnus parviflora* Wall. ex Royle, *Costus speciosus* (Koen.) Sm. and several *Aconitum* spp., belonging to diverse families. While some of these, *e.g. Coptis teeta* Wall., *Podophyllum hexandrum* Royle have been extensively exploited, the others have become rare due to destruction of habitat. Some important medicinal plants are listed in the Table 6.4.

Table 6.4 : Some important plants from N.E. India

Tropical Sub-tropical	Temperate	Sub-alpine and alpine
Abrus precatorius	*Artemisia nilagirica*	*Aconitum chasmanthum*
Acorus calamus	*Berberis asiatica*	*A. deinorrhizum*
Adhatoda zeylanica	*B. wallichiana*	*A. ferox*
Atropa acuminata	*Bergenia ciliata f. ligulatata*	*A. heterophyllum*
Centella asiatica	*Ephedra gerardiana*	*Coptis teeta*
Costus speciosus	*Habenaria commelinifolia*	*Swertia chirayita*
Dioscorea deltoidea	*Hoya globulosa*	*S. hookeri*
Hydnocarpus kurzii	*Gaultheria fragrantissima*	*S. ciliata*
Hyoscyamus niger	*Gentiana kurroo*	*Nardostacnys grandiflora*
Gloriosa superba	*Illicium griffithii*	*Picrorhiza kurrooa*
Mucuna prurita	*Mahonia nepalensis*	*Podophyllum hexandrum*

contd.......

Table 6.4 contd......

Tropical Sub-tropical	Temperate	Sub-alpine and alpine
Ocimum sanctum	*Myrica esculenta*	*Rheum australe*
Paederia foetida		*R. nobile*
Plumbago zeylanica	*Panax pseudo-ginseng*	*Veleriana hardwickii*
Rauvolfia serpentina	*Sarcandra glabra*	*V. jatamansı*
Vetiveria zizanioides	*Saussurea lappa*	

Botanical curiosities

There are some plants in the eastern Himalaya which have created interest among botanists students on account of their special modifications adaptations, and thus add significantly to the floristic diversityof the region. The special features in these plants have endangered some of these plants.

There are some root parasites like *Sapria himalayana* Griff. of the family Rafflesiaceae. This recently discovered largest root parasite has attractive crimson flowers measuring 35 cm across (Deb, 1957). Another root parasite of the family *Mitrastemon yamamotoi* Makino recently collected from Mawsmai forests is a polyendemic. Other root parasites of great scientific interest are *Rhopalocnemis phalloides* Jung. which has been reported from Namdapha Biosphere Reserve in Arunachal Pradesh, *Balanophora dioica* R. Br. and *Boschiniaekia himalaica* Hk. f. & Th. The latter is a root parasite on *Rhododendron* in the alpine areas. *Aeginitia indica* Roxb. is a root parasite on grasses.

There are interesting Saprophytic plants like *Cheilotheca humilis* (D.Don) Keng (*Monotropa uniflora*), *Epipogium roseum* (D. Don). Lindl. and *Galeola falconeri* Hk. f. (Orchidaceae). The latter species is also the tallest orchid in India.

Among insectivorous plants *Nepenthes khasiana* Hk. f. with the leaf tips modified into pitchers measuring up to 12 cm, and two species of *Drosera,* namely *D. peltata* Sm. and *D. burmannii* Vahl are important.

On the other end of the spectrum there are some plants with unusual forms in the high alpine areas. These may be 'cushion forming' or 'snow ball' plants or the 'hot house' plants. Cushion forming plants include species of *Androsace* (Primulaceae), *Saxifraga*

(Saxifragaceae), *Rhodiola* (Crassulaceae), *Thylacospermum* and *Arenaria* (Caryophyllaceae). Several hundred plants aggregate together to form dense, spherical globose cushions. Certainly this is an adaptation agaisnt sever cold and heavy snow fall during witners. One cushion of *Arenaria* or *Thylocospermum* measuring *ca* 30 cm diam takes as much as 100-150 years.

The 'snow ball' plants, *Saussurea gossypiphora* D. Don and *S. graminifolia* Wall. ex DC. (Asteraceae) look like a snow ball due to the dense, white woolly hairs which cover the entire plant and protect from cold wind and snow and keeps warm in day time even if outside temperature suddenly falls. Bees or flies take shelter in the warmth and at the same time pollinate the flowes. The dense woolly hairs that cover the apical meristem act as a sort of thermal insulation (Ohba, 1988).

There is yet, another interesting group of 'hot house' plant like the *Rheum nobile* Hk. f. & Th. and *Saussurea obvallata* (DC.) Edgew., which have their inflorescence sheltered by leafy bracts that can be compared to glasses of a hot 'hot house'. The flower open inside the bracts, where the insects also take shelter for warmth and at the same time pollinate the flowers.

Primitive flowering plants

Eastern Himalaya including north-east India has been considered as a sanctuary of ancient angiosperms. There are a number of primitive flowering plant species which grow in east Himalaya and further eastwards but do not occur in other parts of India (Table 6.5). Takhtajan (1969) based on the analysis of distribution of primitive angiosperms treated the east Himalaya-Fiji region as the 'Cradle of flowering plants', where angiosperms have diversified. East Himalayan flora therefore, has great phytogeographic significance.

North-east Indian rain forests also contribute significantly to the conservation of the world's genetic resources by way of a number of monotypic genera such as *Alcimandra, Aspidocarya, Circaester* and *Hemsleya.* As there are no closely related genomes of these genera anywhere in the world, their conservation is of special significance.

Table 6.5 : Primitive flowering plants occurring in East Himalaya and Notheast India (after Takhtajan, 1969; Rao, 1974)

Species	Family	Distribution
Magnolia griffithii	Magnoliaceae	E. Himalaya and Burma
M. gustavii	-do-	-do-
M. pterocarpa	-do-	-do-
Manglietia sp.	-do-	E. Himalaya, S. China, Indo-China, Java
Euptelea sp.	-do-	E. Himalaya, China, Japan
Tetracentron sinense	Tetracentraceae	E. Himalaya, Burma, S.W. China
Pycnarrhena pleniflora	Menispermaceae	E. Himalaya to N.W. Australia
Haematocarpus thomsonii	-do-	E. Himalaya, W. Malaysia and New Guinea
Aspidocarya uvifera	-do-	E. Himalaya, S.E. Asia
Decaisnea insignis	Lardizabalaceae	E. Himalaya, W. China
Holboellia latifolia	-do-	E. Himalaya, China
Stauntonia spp.	-do-	E. Himalaya, S. China, Taiwan, Vietnam, Korea, Japan
Parvatia brunoniana	-do-	E. Himalaya, S.W. China
Exbucklandia populnea	Hamamelidaceae	E. Himalaya, Sumatra
Distylum indicum	-do-	Himalaya, China, Taiwan, Loas, Korea, Japan
Altingia excelsa	-do-	E. Himalaya, Japan, China, Java, Sumatra
Houttuynia cordata	Piperaceae	Himalaya, China, Japan, Thialand, Taiwan
Myrica esculenta	Myricaceae	E. Himalaya, Korea, Japan, China
Betula alnoides	Betulaceae	Himalaya, E. Asia
Alnus nepalensis	-do-	Himalaya, China

Non-flowering plants diversity

Eastern Himalaya is also well known for the diversity of non-flowering plants. The ferns and fern-allies form a striking feature of vegetation of the eastern Himalaya. Out of about 1000 species of ferns occurring in India 50 per cent are represented in this region. In a recent study of ferns of Meghalaya and Nagaland, about 350 species are reported from these two states alone (Baishya and Rao, 1976; Jamir and Rao, 1986). Some of the rare and interesting ferns of this region are *Dipteris wallichii* (R. Br.) Moore (a polyendemic), *Osmunda cinnamomea* L., *O. claytoniana* L., *O. regalis* L., *Helminthostachys zeylanica* Hk., *Botrychium lanuginosum* Wall., ex Hk. and Grev. var. *lanuginosum*, *Angiopteris evecta* (Forst.) Hoffm., *Cyathea gigantea* (Wall. ex Hk.) Holtt., *C. holttumiana* Rao & Jamir, *C. brunoniana* (Wall.) Cl. & Baker, *Branea insignis* (Hk.) J. Sm. and a few others. The diversity of fern-alies like *Selaginella* and *Lycopodium* is also best expressed in this region (Rao & Hajra, 1986). There are twelve species of *Lycopodium* in E. Himalaya as against only three species in W. Himalaya. Of these, *L. clavatum* L., *L. cernnum* L., L. *serratum* Thunb. and *L. squarrosum* Forst. are quite common.

Unlike pteridophytes, the other lower cryptogams have not received due attention from botanists in our country. Except for the work of Biswas (1934), there is no substantial contribution on the algal flroa of E. Himalaya. Regarding fungi only higher fungi have received some attention (Subramanian and Ramakrishnan, 1952-56). From the work of Awasthi (1965), one can gather an idea of lichen diversity in this region. Several Japanese workers (Asahina and Kurakawa, 1966) have also published on the Lichen flora of E. Himalaya.

Diversity in wild relatives of cultivated plants

The investigations on the centre of origin and the methods of the evolution of cultivated plants have been elaborately discussed by De Candolle (1886) and Vavilov (1926, 52). Vavilov, based on his extensive explorations in different parts of the world, located eight main centres of the origin of crop plants. According to him the north-eastern region of India, forming the 'Hindustani Centre of Origin of Cultivated Plants' is very important for tropical and subtropical fruits, cereals, etc. The north-eastern region forms the richest reservoir of genetic variability of many groups of crop plants. Over 50 species of economic plants have their genetic diversity in this Hindustani Centre of Diversity (Zeven and Zhukovsky, 1975).

Based on geographical distribution, taxonomical and cytogenetical studies, Chakravorty (1951) sugested Assam-Burma-Siam-Indo-Chian region as the centre of origin of *Musa*. This is the area where largest number of species, some of which are endemic, have been recorded. All the basic sets of chromosomes 2n=18, 20, and 24 are found in these species. Banana in north-east India grows wild along the hill slopes of Arunachal Pradesh, Meghalaya and Assam.

This region is also very rich in *Citrus* wealth. There are 64 varieties of *Citrus* which grow wild in this region. Vavilov (l.c.) reported the presence of *Citrus sinensis* (L.) Osbeck, *C. reticulata* Blanc., *C. media* L., *C. aurantium* L., *C. aurantifolia* (Christm.) Swingle and *C. limon* (L.) Burm. f., both as wild and as cultivated forms. The list did not include the Papeda group of *Citrus,* which are very rich in this region. Bhattacharya and Dutta (1956) have described 52 taxa of *Citrus* and 5 probable hybrids from Assam. 'Sch-Niangriang', a wild sweet orange, has been found in Assam (Bhattacharya and Dutta, 1951), which made Tanaka (1958) to suggest that the sweet orange originated in the vicinity of Assam. Further, he concluded that "the presence of *C. indica* Tanaka and its worthy derivatives (*C. depressa* with very large seeds, *C. erythrosa* of Assam Sch-Siem) on the Indian side furnishes a strong evidence that eastern India might be called the centre of origin of all the *Citrus* fruits". To conserve this rich diversity recently a 'Citrus Gene Sanctuary' at Tura, Garo Hills has been established.

Similarly, a rich diversity has been observed in the wild species of temperate fruits, particularly of the family Rosacaeae. *Malus baccata* (L.) Borkh. and *Pyrus pashia* Buch.-Ham. ex D. Don which are used as rootstocks of apple and pear respectively occur in Khasi Hills. Some of the other wild fruit types in this family are *Prunus napaulensis* Steud., *P. undulata* Buch.-Ham. ex D.Don, *P. cerasoides* D. Don, *P. undulata* Buch.-Ham. ex D. Don, *P. cerasoides* D. Don, *P. jenkinsii* Hk.f., *P. wallichii* Steud., *P. persica* Batsch, and *Sorbus khasiana* (Decne.) Rehd. Most of these species occur in Khasi hills and this area perhaps forms the primary/secondary centre of distribution of the genus *Prunus* and *Pyrus.*

Cucurbits, for which E. Himalaya is one of the major centres of origin also exhibits enormous variability in this region. Some of the common cucurbits in this region are : *Lagenaria siceraria* (Mol.) Standl. (bottle gourd), *Luffa cylindrica* (L.) Roem. (sponge gourd), *L. acutangula* (L.) Roxb. (ridge gourd), *Momordica charantia* L. (bitter gourd), *Trichosanthes anguina* L. (snake gourd), *T. Doica* Roxb. (pointed gourd), *Citrullus* sp., *Cucumis melo* L. (long melon), *Benincasa hispida* (Thunb.) Cogn. (ash gourd), *Sechium edule* Sw.

(chayote) and *Cucurbita maxima* Duch., *C. pepo* L. *C. moschata* Duch. (pumpkin). The genus *Cucurbita* has maximum diversity in this region as evident by a number of fruit types that are sold in tribal bazars. The local tribals are also the custodians of the germplasm of several native types of fruits and vegetables including chillies.

Diversity in legume crops and their wild related types in East Himalaya is also of a high order. Maximum variability can be osberved in *Cajanus cajan* (L.) Millsp. (Pigeon pea), *Vigna umbellata* (Thunb.) Ohwi & Ohashi (rice bean), *Vicia faba* L. (broad bean), *Vigna aconitofolia* (Jacq.) Marech. (moth bean), *V. mungo* (L.) Hepper (black gram), *Lathyrus sativus* L. (khesari) and *Glycine max* (L.) Merr. (soya bean). In addition, there are about 200 non-conventional legumes, of which about 50 species are used as vegetables by the various ethnic groups in the region.

It has been estimated that in India about 800 species are consumed as food plants (Singh and Arora, 1978). Of these, about 300 species occur in Eastern Himalaya alone. Arora, Mehra and Nayar (1983), on the basis of the their studies on 323 species of wild relatives of crop plants, observed that rich diversity was concentrated in the warm humid tropical and subtropical regions of north-eastern India.

The region is also unique in having many wild food plants which form the subsidiary food of the local people. Several of these are collected from the wild and marketed in the local tribal and village weekly bazars. Some of the dominant life support species are as follows:

1. *Alternanthera philoxeroides* Griseb. (leaves and young twigs).
2. *Baccaurea ramiflora* Lour. (fruits).
3. *Begonia josephii* A.DC. (leaves).
4. *Bergenia ciliata* (Haw.) Sternb. f. *ligulata* Yeo (leaves)
5. *Calamus erectus* Roxb. (fruits).
6. *Castanea sativa* Mill. (Kernels).
7. *Castonopisis indica* A. DC. (Seeds).
8. *Cinnamomum tamala* Nees (leaves).
9. *Coix lacryma-jobi* L. (seeds).
10. *Colquhounia vestita* Wall. (leaves).

11. *Crassocephalum crepidioides* (Benth.) S. Moore (leaves).
12. *Dillenia indica* L. (fruits).
13. *Dioscorea bulbifera* L. (tubers).
14. *Docynia indica* (Wall.) Decne. (fruits).
15. *Elaeagnus latifolia* L. (fruits).
16. *Eryngium foetidum* L. (leaves).
17. *Eriosema himalaicum* Ohashi (tuberous roots).
18. *Fagopyrum dibotrys* (D. Don) Hara (leaves).
19. *Flemingia procumbens* Roxb. (tuberous roots).
20. *Duchesnea indica* (Andr.) Focke (fruits).
21. *Houttuynia cordata* Thunb. (leaves).
22. *Impatiens longiceps* Hk. f. & Th. (leaves).
23. *Myrica esculenta* Buch.-Ham. (fruits).
24. *Passiflora edulis* Sims. (fruits).
25. *Phoenix humilsi* Royle (fruits).
26. *Emblica officinalis* Gaertn. (fruits).
27. *Polygonum muricatum* Messn. (leaves).
28. *P. orientale* L. (leaves).
29. *P. nepalense* Meissn. (whole plant).
30. *Potentilla fulgens* Wall. ex Hk. (fruits).
31. *Prunus domenstica* L. (fruits).
32. *P. persica* Benth. & Hk. f. (fruits).
33. *P. nepalensis* D. Don (fruits).
34. *Pyrus communis* L. (fruits).
35. *P. pashia* Buch.-Ham. (fruits).
36. *Rubus ellipticus* Sm. (fruits).
37. *R. niveus* Thunb. (fruits).
38. *Rhus semialata* Murr. (fruits).
39. *Viburnum corylifolium* Hk. f. & Th. (fruits).
40. *V. foetidum* Wall. (fruits).
41. *Zanthoxylum acanthopodium* DC. (fruits).

Table 6.6 : Wild relatives of crop plants in different Phytogeographical Zones in India

	Phytogeographical zone	No. of species of wild relatives of crop plants
1.	N.E. region (including Eastern Himalaya)	189
2.	Malabar	142
3.	Western Himalaya	113
4.	Deccan	96
5.	Gangetic Plains	73
6.	North-western plains	49

Recent surveys by I.A.R.I. also have shown that north-east part of India is the richest reservoir for genetic variability in a wide group of crop plants. One of the world's most comprehensive collections of primitive rice cultivars was made by the scientists of I.A.R.I. in localities ranging from 1800-2700 m in Meghalaya and Arunachal Pradesh. Out of 6730 collections of different types of these cultivars, near about 5000 come from the hills. Below 1500 m, the genetic diversity in native banana is enormous and some types show tolerance to diseases, pests and cold conditions. Verma and Ghosh (1979) have made a survey of major *Citrus* growing belts of Meghalaya, Nagaland, Sikkim and Arunachal Pradesh and have described several native varieties of *Citrus*. Many land races of jute and mesta have been found in Tripura, Meghalaya and Garo Hills. Native types of cotton trees are located in the sub-temperate climate of north-eastern hills in Arunachal Pradesh. There are also other fibre yielding plants with many variant forms belonging to genera like *Sesbania, Crotalaria, Sida* and *Cannabis*. According to De Candolle (1886) tea, *Camellia sinensis* (L.) Kuntze was in cultivation in this region for the last 4000 years. The leaves of many allied wild species used as substitute of tea (Maheshwari and Singh, 1964) are found growing here. Hooker (1895) listed 75 species of *Vitis* in India, out of which 46 were found in Eastern India. This suggests the possible origin of this genus in the north-eastern region. Many medicinal plants have been found to be native to this region, *e.g.*, senna (*Cassia angustifolia* Vahl), croton oil (*Croton tiglium* L.), strychnin nut (*Strychnos nuxvomica* L.), chaulmugra oil (*Hydnocarpus kurzii* (King) Warb., mishmi tita (*Coptis teeta* Wall.), rhubarb (*Rheum emodi* Wall. ex Meissn.) and several species of *Dioscorea*.

This region is also regarded as a centre of origin of 5 species of palms of commerce-Coconut (*Cocos nucifera* L.), areca nut (*Areca catechu* L.), palmyra palm (*Borassus flabellifer* L.) sugar palm (*Arenga saccharifera* Labill.) and wild date palm (*Phoenix sylvestris* Roxb.).

It is not possible to list all those wild relatives of cultivated plants which are richly represented here but only a few more important ones are given in the Table 6.7.

Table 6.7 : Some wild relatives of cultivated plants (after Arora, Mehra and Nayar, 1983)

Category of plants	*Name of species*
1. Cereals and millets	*Hordeum agricrithon*
	Digitaria cruciata
	Coix lacryma-jobi
	C. gigantea, C. aquatica
	Oryza rufipogon
	Polytoca wallichiana
2. Legumes	*Moghania vestita*
	M.bracteata
	Vigna capensis
	V. umbellata
	V. pilosa
	Atylosia barbata
	A. scarabaeodes
	A. villosa
	Canavalia ensiformis
	Mucuna bracteata
3. Fruits	*Abelmoschus manihot*
	Duchesnea indica
	Myricaesculenta Prunus acuminata
	P. cerasoides
	P. cornuta
	P. jenkinsii
	P. nepalensis
	Pyrus pashia
	Ribes graciale
	Rubus lineatus
	R. ellipticus
	R. lasiocarpus
	R. moluccanus
	R. reticulatus
	Citrus assamensis
	C. ichangensis
	C. indica

Contd...

Table – *Contd...*

Category of plants	Name of species
	C. jambiri
	C. latipes
	C. macroptera
	C. medica
	C. aurantium
	Docynia indica
	D. hookeriana
	Eriobotrya angustifolia
	Mangifera sylvatica
	Musa acuminata
	M. balbisiana
	M. manii
	M. nagensium
	M. sikkimensis
	M. superba
	M. velutina
	Pyrus pyrifolia
Vegetables	*Cucumis trigonus*
	Luffa graveolens
	Neolufa sikkimensis
	Alocasia macrorhiza
	Amorphophallus bulbifer
	Colocasia esculenta
	Cucumis hystrix
	Dioscorea alata
	Moghania vestita
	Momordica dioica
	M. cochinchinensis
	M. macrophylla
	M. subangulata
	Trichosanthes cucumerina
	T. dioica
	T. dicaelosperma
	T. khasiana
	T. ovata
	T. truncata
	Solanum indicum
5. Oil seeds	*Brassica trilocularis*
6. Fibres	*Corchorus capsularis*
	Gossypium arboreum
7. Species and Condiments	*Allium tuberosum*
	Amomum subulatum

Contd...

Table – *Contd...*

Category of plants	*Name of species*
	Curcuma zedoaria
	Alpinia galanga
	A. speciosa
	Amomum aromaticum
	Curcuma amada
	Piper longum
	P. peepuloides
8. Miscellaneous	*Saccharum longisetosum*
	S. sikkimensis
	S. ravennae
	S. procerum
	S. rufipilum
	Miscanthus nudipus
	M. taylorii
	M. nepalensis
	M. wardii
	Erianthus sp.
	Camellia sp.

Chapter 7

Phyto Sociological Region of the Indian Desert

The Indian Desert lies on the north-western boundary of India and merges with the desert areas of Pakistan, and prior to the separation of these countries was known under a common name "The Great Indian Desert". The Indian Desert lies between 21°25[1] and 30°30′ N and 67° and 75° 25′ E and covers mainly the

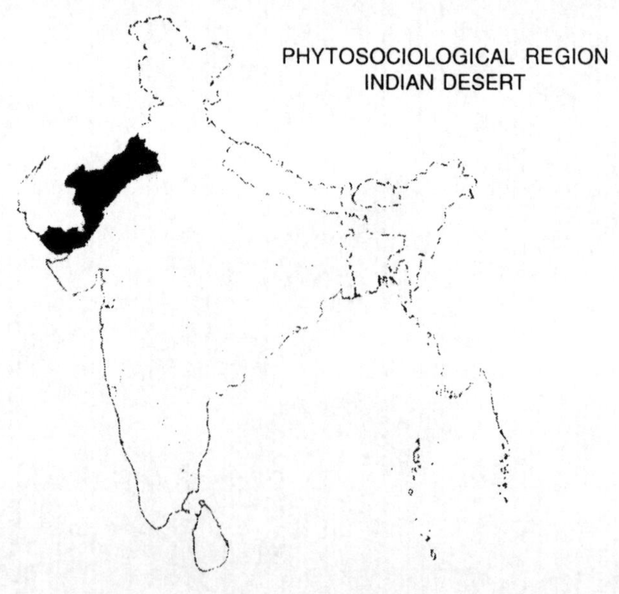

western and north-western region of Rajasthan and a part of the Kutch region of Gujarat in the south-west. Adjacent districts of Haryana and Punjab having extreme arid conditions can also be included. Mainly the districts of Ganganagar, Bikaner, Jaisalmer, Barmer, Jodhpur, Churu and parts of Nagaur, Jalore, Sikar, Pali, Junjhunu districts of north-west Rajasthan and the Rann of Kutch in Gujarat are considered under this phyto sociological region.

The Indian Desert is characterised by high atmospheric temperatue and low and erratic rainfall, high wind velocity, low relative humidity, high evaporation, non-existence of perennial water source and scanty vegetation. The winter (December-January) is quite cold, temperature sometimes falling below freezing point, while the temperature in summer goes up as high as 50°C. The hot season (April-June) is relieved by the south-west monsoon rising from the Arabian sea and the south-east monsoon from the Bay of Bengal (during July-Sept.). The annual rainfall in the desert varies from 88-425 mm and this is too scanty a rainfall to support any appreciable vegetation. The wind velocity in the desert sometimes reaches 140 km per hour from south-west to west.

Vegetational pattern

The vegetation in the desert region is very scanty. The increasing human population and live-stock population is a serious stress on the scanty vegetal cover of the region. Forests occupy only 2 per cent of the geographic area. The botany of Indian desert is fairly well known mainly through the publications of Brandis (1874), Blatter and Hallberg (1918-21), Bhandari (1978), Shah (1978), Majumdar (1979), Shetty and Singh (1987).

Blatter and Hallberg (1918-21) have divided the main types of plant communities as formations, which are exclusively controlled by edaphic factors. The main formations (habitats) based on typical land forms of the Indian desert are

(a) Formation of sand dunes and interdunal areas

This type constitute the major part of the area. The habitat includes sand-dunes (thick deposits of older alluvial but have a clacareous zone at various depths varying from 45-150 cm) of different types, magnitude and orientation. They may be stabilized or unstabilized. The common trees and shrubs of these sand dunes are generally tuft forming bushy types. Important species are *Calligonum polygonoides* L., *Capparis decidua* (Forssk.) Edgew., *Holoxylon salicornicum* (Miq.) Bunge, *Lycium barbatum* L., *Ziziphus nummularia* (Burm. f.) Wt. & Arn, *Acaicai senegal* (L.) Willd. *Prosopis cineraria* (L.) Druce, *Salvadora oleoides* Decne. The other low herbaceous and undershrubs are *Aerva javanica* (Burm. f.) Juss. ex Schult., *Citrullus colocynthis* (L.) Schrad., *Crotalaria burhia* Buch.-Ham., *Dipterygium glaucum* Decne., *Indigofera argentea* Burm., f., *I. cordifolia* Heyne ex Roth. *I. linifolia* (L.f.) Retz., *Leptadenia pyrotechnica* (Forssk.) Decne., *Melhania denhamii* R. Br., *Sericostemma pauciflorum* Stocks and *Tribulus longipetalus* Viv. Often, there are characteristic associations of *Crotalaria burhia* Buch.-Ham. *Aerva* spp. and *Leptadenia pyrotechnica* (Forssk.) Decne. on low sandy dunes and sandy plains.

The interdunal gaps support comparatively a luxuriant vegetation due to greater availability of moisture. Common species in such areas are *Acacia jacquemontii* Benth., *A. senegal* (L.) Wild., *Prosopis cineraria* (L.) Druce, *Salvadora oleoides* Decne., *Tecomella undulata* (Sm.) Seem., *Capparis decidua* (Forssk.) Edgew., *Zizphus nummularia* (Burm. f.) Wt. & Arn. These trees and shrubs often provide space for the growth of tufts of *Convolvulus microphyllus* Sieb. ex Spreng., *Polygala irregularis* Boiss., *Tephrosia purpurea* (L.) Pers., *Cenchrus biflorus* Hk.f., *Mimosa hamata* Willd., etc. The bush-forming *Acacia jacquemontii* Benth., *Capparis decidua* (Forssk.) Edgew. and *Calligonum polygonoides* L. are able to survive the invasion of shifting sand dunes.

(b) Vegetation of sandy and hammocky plains

Most of the species in this habitat possess well developed root system and occur in open clump formations, which provide suitable habitat for some of the rainy season species such as grasses of the genera *Aristida, Cenchrus, Panicum* and some sedges.

The common trees and shrubs in this type of vegetation are *Acacia senegal* (L.) Willd., *Calligonum polygonoides* L., *Capparis decidua* (Forssk.) Edgew., *Maytenus senegalensis* (Lam.) Exell., *Prosopis cineraria* (L.) Druce, *Salvadora oleoides* Decne., *S. persica* L. and *Ziziphus nummularia* (Burm. f.) Wt. & Arn. The herbs and low shrubs are constituted by *Aerva javanica* (Burm. f.) Juss. ex Schult., *Arnebia hispidissima* DC., *Boerhavia diffusa* L., *Convolvulus microphyllus* Sieb.

ex Spreng., *Crotalaria burhia* Buch.-Ham. and *Leptadenia pyrotechnica* (Forssk.) Decne. The common creepers mostly belong to the Cucurbitaceae family. These are *Citrullus colocynthis* (L.) Schrad., *C. lanatus* (Thunb.) Mats. & Nakai, *Cucumis melo* Cogn., *Coccinia grandis* (L.) Voigt, *Momordica dioica* Roxb. ex Willd.., *Mukia maderaspatana* (L.) Roem. and the Asclepidaceae member *Pergularia daemia* (Forssk.) Chiov. The shifting dunes are often successful in over running the low vegetation.

(c) Gravelly and rocky plains and isolated hills

The species coming up on these gravelly plains are altogether different compared to the vegetation of sand-dunes. Some of the common plants of these habitats are *Cleome vahliana* Farsen., *C. scaposa* DC., *Dactyloctenium aristatum* Link, *Indigofera linnaei* Ali, *Heliotropium rariflorum* Stocks., *Blepharis sindica* Stocks., *Sericostemma pauciflorum* Stocks, *Bouchnera marubifolia* Schauer and *Salvia aegyptiaca* L. Some species like *Euphorbia clarkeana* Hk.f., *E. granulata* Forssk., *Indigofera cordifolia* Heyne ex Roth, *Mollugo cerviana* (L.) Ser., *Tribulus terrestris* L., etc. have star-like prostrate branches remaining appressed to the ground. The trees and shrubs in this habitat are *Calotropis procera* (Ait.) R.Br., *Capparis decidua* (Forssk.) Edgew., *Euphorbia caducifolia* Haines, *Maytenus senegalensis* (Lam.) Exell., etc. Most of these plants are often covered by ramblers like *Coccinia grandis* (L.) Voigt, *Dactyliandia welwitschii* Hk.f. and *Pergularia daemia* (Forssk.) Chiov. These gravelly plains also support grass-legume association of *Enneapogon-Tragus-Indigofera* spp.

Near the foothills *Ephedra foliata* Boiss. & Kotschy ex Boiss., *Asparagus racemosus* Willd. and grasses of the genera *Eleusine, Aristida, Oropetium* and *Melanocenchris* form the common associates.

The rocky plains have very sparse vegetation mainly of *Euphorbia caducifolia* Haines, *Anogeissus* pendula Edgew., *Balanites aegyptiaca* (L.) Del., *Corallocarpus epigaeus* (Rottl. ex Willd.) Cl., *Rivea hypocrateriformis* (Desr.) Choisy, and *Rhynchosia minima* (Camb.) Baker. The undergrowth is predominated by the species of *Blepharis, Barleria, Cleome, Fagonia* and a few others. Increasing density of these species can be observed at the foot hills.

Vegetation of Saline tractrs

There are large saline tracts spread throughout the desert extending to the Rann of Kutch in Gujarat. Kutch is practically an undulating rocky area with a Great Rann of Kutch at the north and east and the Little Rann on the south-east consisitng of vast tidal mud flats with saline efflorescence. The Rann is a dry bed of the remnant of an arm of the sea which once connected the Narmada rift with the Sindh and separated Kutch from the mainland. Almost from November to March the Rann is a barren tract of dry salt encrusted mud, presenting aspects of almost inconceivable desolation.

Although Kutch Peninsula containing the Great Rann and the Little Rann of Kutch are excluded from the Desert Biogeographic province by several workers, the present author includes this in Desert Biogeographic province mainly because of the vegetation which is so typical of the desert region. Many species recorded here also prevail in the desert zone.

The vegetation here consists of typical salt marsh-salt bush plant community of halophytes. In the interior, where the soil is less sandy and saline, a number of plants typical of the habitat are seen. Some such species are *Cressa cretica* L., *Haloxylon* recurvum (Moq.) Bunge ex Boiss., *H. salicornicum* (Moq.) Bunge, *Salsola baryosma* (Roem. & Schult.) Dandy, *Sesuvium sesuvioides* (Fenzl.) Verdc., *Suaeda fruticosa* (L.) Forssk., *Tamarix indica* Willd., *Zygophyllum simplex* L., *Salvadora persica* L., *Atriplex stocksii* Boiss., *Capparis decidua* (Forssk.) Edgew., *Scirpus tuberosus* Desf., *Malachra capitata* L., *Lycium barbatum* L., *Fimbristylis* spp., *Jucus maritimus* L., *Spinifex littoreus* (Burm. f.) Merr. and a few others.

A number of species also ocur in the sandy belt of Kutch, many of which are also the common elements in the desert zone of Rajasthan. These are *Arnebia hispidissima* DC., *Crotalaria burhia* Buch.-Ham. ex Benth., *Farsetia jacquemontii* Hk.f. & Th., *Leptadenia pyrotechnica* (Forssk.) Decne., *Pedalium murex* L., *Peganum harmala* L., *Seetzenia lanata* (Willd.) Bullock., *Sericostoma pauciflorum* Stock. ex Wt., *Tribulus longipetalus* Viv., *Zygophyllum simplex* L., *Polycarpaea* corymbosa (L.) Lam., *Portulaca pilosa* L. and a number of grasses mainly belonging to *Aristida hystrix* L.f., *Cenchrus biflorus* Roxb., *Chloris montana* Roxb., *Desmostachys bipinnata* (L.) Stapf,

Melanocenchris jacquemontii Jaub. & Spach., *Tragus roxburghii* Panigr., *Sehima nervosum* (Rottl.) Stapf. *Urochondra setulosa* (Trin.) C.E. Hubb.

The Indian Desert phyto sociological region is a store house of genetic resources of a number of life support species, typical of the arid and semi-arid zone. Fruits, leaves, grains and tubers from several species are consumed by the local people as subsidiary food during scarcity of food. Remarkable diversity is also exhibited by these groups of plants. A few important plant species used as leafy vegetables are *Amaranthus viridis* L. 'chandlai', *A. spinosus* L. 'jangli chandlai', *A. hybridus* L. 'lal cholai', *Achyranthes aspera* L. 'unda kante', *Acalypha indica* L. 'khokte', *Portulaca oleracea* L. 'jangli kulfa', *Gisekia pharnaceoides* L. 'morang, *Celosia argentea* L. 'murga kalgi', *Digera muricata* (L.) Mart. 'morang', *Portulaca quadrifida* L. 'luni or lunki', *Cleome viscosa* L. 'bagre', *Commelina benghalensis* L. 'bakna', *C. forsskalii* Vahl 'kansura', *Cassia occidentalis* L. C., *obtusa* Roxb., 'kesudo', *Leucas aspera* Spreng. 'goma', *L. cephalotus* Spreng. 'halkhura', *Ocimum americanum* L., *O. basilicum* L. 'bapchi' *Euphorbia caducifolia* Haines 'thor', *Securnega leucopyrus* Brandis 'arg', *Chenopodium album* L. 'bathua', *C. murale* L. 'goila', *Launaea pinnatifida* Cass. 'pathri', *Ipomoea aquatica* Forssk. 'narz', *Cocculus pendulus* (J.R. & G. Forst.) Diels 'peelwan', *Cardiospermum halicacabum* (L.) 'kanphuti', *Boerhavia diffusa* L. 'sata', *B. verticillata* Poir. 'mota sata', *Bambusa arundinaceae* Willd. 'bamboos', *Moringa oleifera* 'saijna', and *Tamarindus indica* L. 'imli'.

Similarly, seeds of grasses viz., *Cenchrus* spp., *Panicum* spp., *Brachiaria* spp., *Lasiurus* spp., *Eleusine* spp., *Dactyloctenium* spp., 'pearl millet', *Citrullus lanatus* 'mateera', *Acacia senegal* (L.) Willd., 'kumut', *Indigofera* spp., *Vigna trilobata* (L.) Verdc., Cenchrus ciliaris L. 'dhaman', *C. setigerus* Vahl 'motha dhaman', *C. biflorus* Roxb. 'bhurat', *Lasiurus scindicus* Stapf 'sewan', *Panicum antidotale* Retz. 'gramma', *P. turgidum* Forssk. 'murut', *Brachiaria ramosa* (L.) Stapf 'kuri', *Dactyloctenium scindicum* Boiss. 'ganghi', *D. aegypticum* (L.) P. Beauv. 'makta', *Eleusine compressa* (Forssk.) Aschers. & Schweinf. 'tantia', *Echinochloa colona* (L.) Link 'soma', *E. crusgalli* (L.) P. Beauv. 'samaw', *Vigna trilobata* (L.) Verdc., *Indigofera* spp., *Ocimum americanum* L. 'bapchi', *O. basilicum* L.f. 'meruva', *Sesbania aegyptiaca* Pers. 'ekad', *Abutilon indicum* G. Don 'tara kanchi', *Acacia nilotica* (L.) Willd. ex Del. 'babool', *A.leucophloea* (Roxb.) Willd. 'urojio', and *Tamarindus indica* L. 'imli' are also eaten.

The following species yield edible fruits :

Ziziphus nummularia (Burm. f.) Wt. & Arn 'bordi', *Capparis decidua* (Forssk.) Edgew. 'kair', *Rhus mysorensis* Heyne 'dansi', *Grewia tenax* (Forssk.) Ashcers. & Schweinf. 'gengeren', *G. villosa* Willd. gangeti', *Opuntia ficus-indica* (L.) Mill. 'thor', *Cordia myxa* Roxb. 'gunda', *Moringa oleifera* Lam. 'saijna', *Acacia senegal* Willd. 'kumut', *Prosopis cinerarai* (L.) Druce 'khejri, sangri', *Grewia asiatica* L. 'falsa', *Calligonum polygonoides* L. 'phog', *Ficus benghalensis* L. 'barga', *F. religiosa* L. 'pipal', *F. racemosa* L. 'gular', *Salvadora oleoides* Decne. and *S. persica* L. 'jal', *Pithecellobium dulce* Benth. 'jangal jalebi', *Cucumis callosus* (Rottl.) Cogn. 'kachra machri' and *Momordica* spp. 'kankera, bankarela'.

Flowers of *Calligonum polygonoides* L. 'phog', *Moringa oleifera* Lam. 'saijna', *Prosopis cineraria* (L.) Druce 'khejri', are also eaten as vegetable.

Roots and tubers of *Chlorophytum* sp. 'safed mushli', bulbs of *Ceropegia tuberosa* Roxb. 'khadula', rhizomes of *Cyperus* spp., roots of *Asparagus racemosus* Willd. 'satavar', *Butea monosperma* (Lam.) Taub. 'dhak' and *Bombax ceiba* L. 'semul' are edible. Leaf pulp of *Aloe vera* L. 'guar patha', bark of khejri are also edible.

The genetic resources and diversity needs to be conserved in the following endangered species. *Citrullus colocynthis* (L.) Schrad., *Withania coagulans* (Stocks.) Dunal, *Commiphora wightii* (Arn.) Bhandari, *Caralluma edulis* (Edgew.) Benth. & Hk.f., *Dipcadi erythraeum* Benth., *Farsetia jacquemontii* Hk.f. & Th., *Sida tiagi* Bhandari, *Tribulus rajasthanensis* Bhandari & Sharma, *Monsonia beliotropioides* (Can.) Boiss., *Tephrosia falciformis* Ramas., *Psoralea odorata* Blatt. & Hallb., *Pulicaria rajputanae* Blatt. & Hallb., *Odontanthera varians* (Stocks.). Mabberley & Hk.f., *Convolvulus stocksii* Boiss., *Anticharia glandulosa* Aschers. var. *Caerulea* Blatt. & Hallb., *Cenchrus rajasthanensis* Kanodia & Nanda. Their populations have already reached a critical level due to species fragmentation.

Chapter 8

Phyto Sociological Region of Semi-Arid Zone

Semi-arid is a region of transition between the true desert in the west to extensive Deccan communities of Peninsular India in the south and east. Therefore, the boundaries are not sharp and absolute, but a gradual continuum. Mainly the area surrounding the Desert zone of western Gujarat and Rajasthan are considered

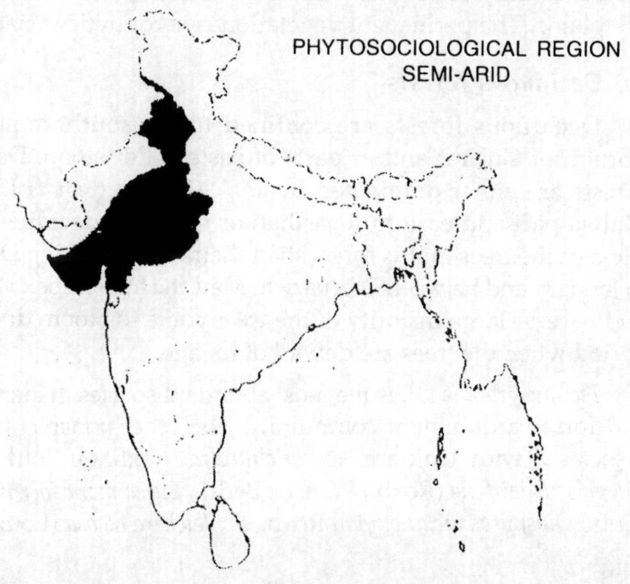

under this (The Punjab plains, including Delhi, Haryana, fringes of Jammu & Kashmir, Himachal Pradesh and western edge of Uttar Pradesh and eastern Rajasthan, eastern Gujarat and north-west Madhya Pradesh).

This zone has strong biological links with western Asia and northern Africa. Many of the taxa characteristic of this zone reveal African affinity, i.e. *Acacia, Anogeissus, Balanites, Capparis* and *Grewia.*

The climate is very hot in summer and markedly cold in winter. April to June are the hotter months with temperature rising up to 48°C in May and June. October-January is the winter season with January being the coldest month when temperature falls down to as low as 2°C and sometimes below the freezing point. The rainfall is very low, the average annual rainfall varies between 400-1000 mm, the Siwalik region receiving the heaviest monsoon. July is the rainiest month. Dust storms are also frequent during March-April.

The soils in the plains consist of Indo-Gangetic alluvium. The soils vary for red loam to black cotton clay in the south-western and southern parts of the zone.

Vegetation

The vegetation in the semi-arid region varies from (a) deciduous forests in the isolated hilly portions to (b) open scrub jungles in the plains. The permanent vegetation is xerophytic in nature.

(a) Deciduous forests

Deciduous forests are confined to the southernmost hilly portion of Gujarat and in parts of eastern Rajasthan. Deciduous forests are either domianted by Teak (*Tectona grandis* L.f.) as in Malwa part and eastern Rajasthan or non-Teak deciduous types. Most of the trees in this forest shed their foliage during December to January and between February to May, the forests look very open and bare. A large majority of the trees come to bloom during this period when the trees are devoid of foliage.

Tectona grandis L.f. is the most abundant species in many places and forms a dominant community. The other conspicuous trees associated with Teak are *Acacia chundra* (Roxb. ex Rottl.) Willd., *Anogeissus latifolia* (Roxb.) Wall. ex Bedd., *Butea monosperma* (Lam.) Taub., *Diospyros melanoxylon* Roxb., *Schleichera oleosa* (Lour.) Oken,

Lannea coromandelica (Houtt.) Merr. and *Boswellia serrata* Roxb. ex Colebr. Apart from the above dominant species, deciduous forests where Teak is not dominant, also contain *Haldina cordifolia* (Roxb.) Ridsd., *Albizia lebbeck* (L.) Benth., *A. odoratissima* (L.f.) Benth., *Bombax ceiba* L., *Bauhinia racemosa* Lam., *Careya arborea* Roxb., *Emblica officinalis* Gaertn., *Cochlospermum religiosum* (L.) Alst., *Gmelina arborea* L. *Lagerstroemia parviflora* Roxb., *Wrightia tinctoria* R. Br., *Morinda pubescens* Sm., *Sterculia urens* Roxb. and rarely *Ougenia oogeinensis* (Roxb.) Hochst. In some places the forests are dominated by *Anogeissus-Boswellia* association. *Dendrocalamus strictus* (Roxb.) Nees comes up in comparatively open places forming huge clumps.

The Aravalli range with Mount Abu as the highest peak (1727 m) presents a diferent pattern of plant distribution. The top of the small hills (less than 600 m) on this range are barren, while on the slopes, where some soil accumulates, dense growth of *Acacia leucophloea* (Roxb.) Willd., *A. senegal* Willd., *Balanites aegyptiaca* (L.) Del., *Capparis decidua* (Forssk.) Edgew., *Euphorbia nivulia* Buch.-Ham., *Securinega leucopyrus* Muell., etc., are noticed. The hills above 600 m are clothed by trees such as *Anogeissus latifolia* (Roxb.) Wall. ex Bedd., *A. pendula* Edgew., *Balanites aegyptiaca* (L.) Del., *Prosopis cineraria* (L.) Druce, *Wrightia arborea* (Dennst.) Mabberley, *Commiphora wightii* (Arn.) Bhandari, *Bauhinia racemosa* Lam., and *Sterculia urens* Roxb., all of which represent the typical elements of deciduous forests.

Mount Abu shows distinct elevational zones. The chief tree components up to 1300 m altitude are *Anogeissus pendula* Edgew., *Aegle marmelos* (L.) Corr., *Boswellia serrata* Roxb. ex Colebr., *Cassia fistula* L. *Mitragyna parvifolia* (Roxb.) Korth., *Diospyros melanoxylon* Roxb. and *Wrightia tinctoria* R. Br. These are essentially the same species as those of deciduous forest. Above 1300 m, contrary to the semi-arid type a subtropical evergreen type of vegetation is met with.

Although no stratification can be marked in these forests, some smaller trees and shrubs, though scattered, are worth mentioning. These are *Wrightia tinctoria* R. Br., *Alangium salviifolium* (L.f.) Wang., *Crissa congesta* Wt., *Mallotus philippensis* (Lam.) Muell.-Arg., *Trema orientalis* (L.) Bl., and *Ziziphus xylopyra* (Retz.) Willd. The other shrubs in the forest composition are *Xeromphis spinosa* (Thunb.) Keay, *Meyna laxiflora* Robyns., *Flacourtia indica* (Burm. f.) Merr.,

Azanza lampas (Cav.) Alef., *Barleria cristata* L., *Helicteres isora* L., *Hibiscus caesius* Garcke, *Fioria vitifolia* (L.) Mattei. and *Ficus exasperata* Wall.

On the outskirts of forests, where trees are scarce, shrubs like *Capparis sepiaria* L., *C. decidua* (Forssk.) Edgew., *Maytenus senegalensis* (Lam.) Exell., *Mimosa hamata* Willd., *Maerua oblongifolia* A. Rich, *Adhatoda zeylanica* Medic., *Cassia auriculata* L. and a host of others form the striking feature.

The notable climbers/stragglers in these forests are *Aristolochia indica* L., *Cryptolepis buchanani* Roem. & Schult., *Gymnema sylvestre* (Retz.) R. Br., ex Schult., *Cissus repanda* Vahl, *Cansjera rheedii* Gmel., *Celastrus paniculatus* Willd., *Combretum albidum* G. Don, *Hiptage bengalensis* (L.) Kurz, *Ventilago denticulata* Willd., *Butea parviflora* Roxb., *Asparagus racemosus* Willd., *Wattakaka volubilis* (L.f.) Stapf and *Abrus precatorius* L.

The herbaceous vegetation is predominantly of the monsoonic type. These are found either as scattered individuals or forming dense populations. The common ones are *Acanthospermum hispidum* DC., *Amaranthus spinosus* L., *Achyranthes aspera* L., *Blainvillea acmella* (L.) Philipson, *Cassia absus* L., *C. occidentalis* L., *C., tora* L., *Cleome gynandra* L., *C. viscosa* L., *Martynia annua* L., *Sida* spp., *Desmodium* spp. and in completely open places grass-legume association of *Arthraxon-Dichanthium - Zornia or Eragrostis - Panicum Desmodium* can be noticed.

In many places, these thorny scrubs form impenetrable thickets and under these thickets, where there is accumulation of some amount of humus and moisture, shade-loving plants like *Oxalis corniculata* L., *Biophytum sensitivum* (L.) DC., *Lindernia ciliata* (Colsm.) Pennell, etc., are common.

(b) Open scrub jungles

The open scrub jungles are quite predominant in the entire semi-arid tracts. The species composition more or less remains the same throughout the scrub jungles in Inda, except for the species density which varies from place to place. For example, in Kathiawar *Acacia-Capparis* scrublands are dominant, while in Aravalis *Acacia-Anogeissus pendula* forests are dominant type. The vast majority of the species are drought resistant. The tree species if any are widely scattered.

The typical members of the scrub vegetation are *Acacia senegal* Willd., *A. farnesiana* (L.) Willd., *A. jacquemontii* Benth., *Balanites aegyptiaca* (L.) Del., *Capparis decidua* (Forssk.) Edgew., *C. sepiaria* L., *C. zeylanica* L., *Dichrostachys cinerea* Wt. & Arn., *Commiphora wightii* (Arn.) Bhandari, *Limonia acidissima* L., *Ziziphus nummularia* (Burm.f.) Wt. & Arn., *Z. horrida* Roth, *Z. mauritiana* Lam., *Kirganelia reticulata* (Poir.) Baill., *Mimosa hamata* Willd., *Cassia auriculata* L., *Salvadora persica* L., *Prosopis cineraria* L., *Caesalpinia bonduc* (L.) Roxb. and a few others.

The scattered tree species in these scrub jungles are *Acacia leucophloea* (Roxb.) Willld., *A. nilotica* (L.) Del. ssp. *indica* (Benth.) Brenan, *Aegle marmelos* (L.) Corr., *Butea monosperma* (Lam.) Taub., *Cassia fistula* L., *Mangifera indica* L., *Diospyros melanoxylon* Roxb., *Cordia dichotoma* Forst. f., *Phoenix sylvestris* (L.) Roxb., *Azadirachta indica* A. Juss., *Miliusa tomentosa* (Roxb.) Sincl., *Streblus asper* Lour. and *Ficus* spp. Many of these thorny shrubs and trees frequently get infested with a number of stem parasites like *Cassytha filiformis* L., *Cuscuta chinensis* Roxb., *Dendrophthoe falcata* (L.f.) Etting and *Viscum articulatum* Burm.f.

The climbers on these thorny bushes and small trees belong to diverse families. Some of the important ones are *Cryptolepis buchananii* Roem. ex Schult., *Gymnema sylvestre* (Retz.) R. Br. ex Schult., *Holostemma ada-kodien* Schult., *Tylophora hirsuta* (Wall.) Wt. & Arn., all belonging to Asclepiadaceae; *abrus precatorisus* L., *Mucuna pruriens* (L.) DC., *Canavalia gladiata* (Jacq.) DC., of the Fabaceae; *Cissampelos pariera* L. (Menispermaceae); *Aristolochia indica* L. (Aristolochiaceae); *Dioscorea hispida* Dennst., *D. bulbifera* L. (Dioscoreaceae); *Gloriosa superba* L. (Liliaceae) and *Cardiospermum halicacabum* L. (Sapindaceae).

The herbaceous flroa is also very rich and profuse during rainy season and belong to varied families such as Asteraceae, Caesalpiniaceae, Amaranthaceae, Euphorbiaceae, Lamiaceae, Tiliaceae, Aizoaceae and Poaceae. The open grassland vegetation includes species of *Aristida adscensionis* L., *Bothriochloa pertusa* (L.) A. Camus, *Cenchrus ciliaris* L., *Chloris inflata* Link, *Eragrostis unioloides* (Retz.) Nees ex Steud., *Brachiaria ramosa* (L.) Stapf, *Echinochloa colonum* (L.) Link, *Heteropogon contortus* (L.) P. Beauv., *Dichanthium caricosum* (L.) A. Camus, *Digitaria ciliaris* (Retz.) Koel.,

Desmostachya bipinnata (L.) Stapf, *Themeda quadrivalvis* (L.) Kuntze and *Sehima nervosum* (Hottl. ex Willd.) Stapf. In these grasslands root parasites like *Buchnera hispida* Buch.-Ham., *Orobanche aegyptiaca* L., *O. cernua* Loefl., *Sopubia delphinifolia* (Roxb.) G. Don, *S. trifida* Buch.-Ham. ex G. Don and *Striga* spp. are also frequently noticed.

Apart from the above two major types of vegetation, hydrophytic and marsh flora in this zone also reveals much diversity. But as most of these aquatic species are common to almost all the biogeographic zones of India, these are dealt in a separate section.

Chapter 9

PHYTO SOCIOLOGICAL REGION OF GANGETIC PLAINS

The Gangetic plain phyto sociological region stretching from eastern Rajasthan through Uttar Pradesh to Bihar and West Bengal is mostly under agriculture and supports dense human population. The northern limit is marked by the Siwalik hills; in south the limits are not sharp, but the Vindhyan escarpment and northern and

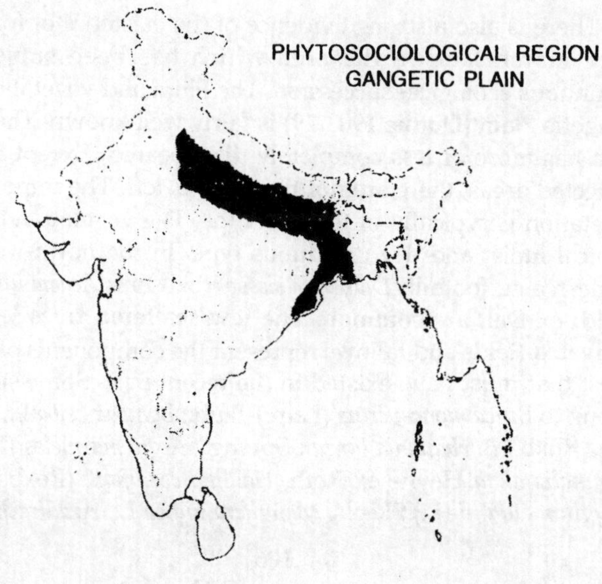

eastern outliers of the Chota Nagpur and Aravali range can be considered the southern boundary. The entire region is a flat alluvial region. The Siwalik hills are not considered under this zone, although Rodgers and Panwar (1988) do so. The Sunderbans, having the salt water components of the estuaries and deltas which should have normally formed the part of the Gangetic plain are discussed under the coastal zone.

The Gangetic plain goes from great aridity in the west, below 500 mm rainfall to great moist conditions in the east with above 4000 mm annual rainfall.

Vegetation

Being the most densely populated region of India and having been under continuous and intensive cultivation, the area has lost its original flora, except perhaps the *Shorea* forest in the Terai region. Till the late 16th century the area supported dense monsoon deciduous forests sheltering a number of wild animals. In fact, the area surrounding Delhi formed a famous hunting ground of the Moghals.

The secondary vegetation in abandoned fallows is marked by xerophytic flora. Tremendous increase in human population in recent years has further replaced even the xerophytic vegetation to open dry grasslands, forming grazing tracts.

There is also a strong evidence of the existence of formerly a large marsh land in the area which has been subjected to continuous ecological succession. The Flora and vegetation of the Gangetic Plain (Duthie 1903-29) is fairly well known. The natural flora has more or less completely disappeared. Except for some protected areas, there are not true forests left. The impoverished vegetation is typically of the arid type. The vegetation is chiefly tropical moist and dry deciduous type. In the north-west Uttar Pradesh near foothills *Dalbergia sissoo* Roxb. and *Acacia nilotica* (L.) Willd. ex Del. are common. The few scattered trees left in the cultivated fields and fallows represent the components of the true forest that must have existed in the recent past. Some such trees belong to *Butea monosperma* (Lam.) Taub., *Bombax ceiba* L., *Sterculia urens* Roxb., *Buchanania lanzan* Spreng., *Aegle marmelos* (L.) Corr., *Terminalia alata* Heyne ex Roth, *Haldina cordifolia* (Roxb.) Ridsd., *Syzygium cumini* (L.) Skeels, *Mangifera indica* L., *Azadirachta indica*

Juss., *Mallotus philippensis* Muell. – Arg., and *Lagerstroemia parviflora* Roxb., all forming the components of the deciduous forests.

In the extreme western part of the Gangetic plain, some species typical of the Indian plain such as *Balanites aegyptiaca* (L.) Del., *Boswellia serrata* Roxb., *Lannea coromandelica* (Houtt.) Merr., *Prosopis cineraria* (L.) Druce, *Anogeissus pendula* Edgew., *Acacia* spp., *Cordia dichotoma* Forst., *Olea ferruginea* Royle and *Ziziphus* spp. are common. Shrubs like *Flacourtia indica* (Burm.f.) Merr., *Capparis* spp., *Cassia auriculata* L., *Dodonaea viscosa* L., *Calotropis procera* R.Br., *Withania coagulans* Dunal, *Adathoda zeylanica* Medic. are common.

The Gangetic Plain biogeography is also characterised by having tall grasses of the genera *Themeda, Saccharum, Narenga, Phragmites, Vetiveria* and several others. There are also large conspicuous patches of species of *Chloris, Cymbopogon, Apluda, Bothriochloa, Dichanthium, Desmostachya, Panicum, Cynodon, Andropogon,* etc. which form the major fodder resource of the region. Due to the increased human activity, the weed flora has dominated the landscape. The major weeds belong to *Xanthium strumarium* L., *Parthenium hysterophorus* L., *Peristrophe bicalyculata* (Retz.) Nees, *Amaranthus spinosus* L., *Lantana camara* L., *Cassia tora* L., *Chenopodium* spp. and waste places *Cannabis sativa* L. (bhang).

The south-eastern end of the Gangetic Plain merges with the littoral and mangrove region of Sunderbans and the latter has been discussed under the Coastal Biogeographic zone.

Chapter 10

Phyto Sociological Region of the Western Ghats

The Western Ghats phyto sociological region which forms the "Malabar" province according to several phytogeographers (Hooker, 1904; Clarke, 1898; Chatterjee, 1940) is a narrow stretch running from the hills south of Tapati river in the north to Kanyakumari in the south, along the west coast of India. The zone

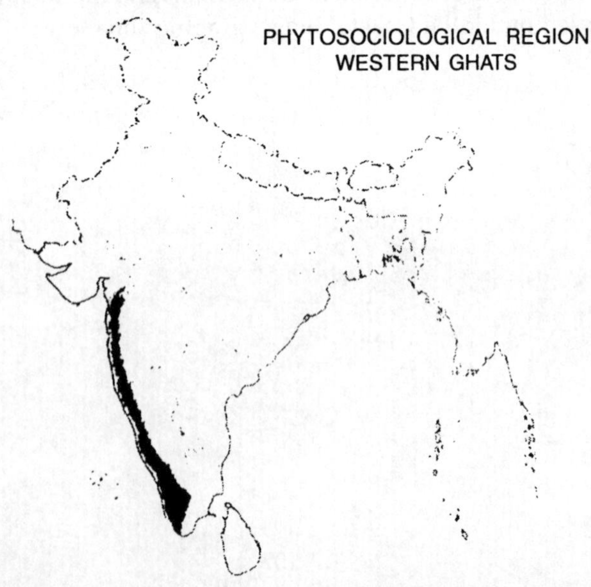

lies between 22° N to 8°N and runs approximately for 1500 km, encompassing a considerable gradient of climatic conditions. The Ghats rise up abruptly in the west to a highly dissected plateau up to 2900 m in height and descend to the dry Deccan plains below 500 m in the east. The highest point is Doddabett (2920 m) in the Nilgiris. The rainfall pattern sia slo extremely variable from 5000 mm per annum in windward areas to under 600 mm in the leeward or rain-shadow areas with prolonged dry season.

The climatic and altitudinal gradient has resulted in a vareity of forest types–from evergreen to semi-evergreen; from moist deciduous to dry deciduous formations. In the higher hills stunted montane communities have also developed.

Vegetation

Our knowledge of the vegetation of Western Ghats region is derived largely from the works of Gamle (1915-36), Puri (1960), Fyson (1932), Subramanyam and Nayar (1974), Rao (1985), Nair and Daniel (1986) and a few others. The Western Ghat is one of the major tropical evergreen forested regions in India and exhibits and enormous plant diversity. As many as 4000 species of flowering plants are recorded from this region, of which more than 50 per cent are endemic to this region. Another interesting aspect of the flora is the preponderence of species of Clusiaceae, Dipterocarpaceae, Myristicaceae, Sterculiaceae, Myrtaceae, Melastomataceae, Gesneriaceae and Pipercacea. The conifer, *Podocarpus latifolia* Wall. is confined to the hills in Tinervelly in the whole of Peninsula and outside it is known only from Burma and Malaya. *Bentinckia conpanna* Berry, is another remarkable species of Travancore hills.

Based on the floristic composition and other natural factors, the Western Ghats have ben divided into 4 phytogeographyical units viz. (a) the Western Ghats from river Tapati to Goa, (b) from river Kalinadi to Coorg, (c) the Nilgiri and (d) the Anamalai, Palni and Cardamon hill. But, for the purpose of present discussion the Western Ghats as a whole is treated as one zone, as more or less the forest types remain the same in all the 4 regions. Mainly the following forests types occur in the Western Ghats.

(i) The dry scrub vegetation
(ii) Dry deciduous forests
(iii) Moist deciduous forests
(iv) Semi-evergreen forests
(v) Evergreen forests
(vi) The 'Sholas'
(vii) Grasslands.

(i) The dry scrub vegetation

The dry scrub vegetation or the semi-desert vegetation occurs at the foothills, particularly along the eastern side of the Western Ghats. The elevation varies from 200 to 400 m and the rainfall ranges from 350 to 6000 mm. The vegetation is mostly comprised of thronyspecies. Some of the common ones are *Barleria prionitis* L., *B. cristata* L., *Eranthemum roseum* R.Br., *Hemigraphis latebrosa* Nees, *Rungia repens* Nees, *Dicliptera foetida* (Forssk.) Blatt., *Aerva sanguinolenta* Bl., *Mimosa pudica* L., *Corchorus trilocularis* L., *Jatropha curcas* L. and species of *Acacia.* As one proceeds southwards, the species composition of the vegetation slightly changes. Here, the following species are conspicous. *Commiphora berryi* Engl., *Dichrostachys cinerea* Wt. & Arn., *Acacia horrida* (L.) Willd, *A. planifrons* Wt. & Arn., *Opuntia dillenii* Haw., *Dichoma tomentosa* Cass., *Azima tetracantha* Lam., *Solanum triblobatum* L., *Euphorbia antiquorum* L., *Dodonaea viscosa* L., *Flacourtia indica* (Burm. f.) Merr., *Carissa congesta* Wt., *Capparis* spp., *Xeromphis spinosa* (Thunb.) Keay, *Rhus mysorensis* Heyne, *Erythroxlum monogynum* Roxb. and *Balanites aegyptiaca* (L.) Del.

The tree species are very few and are mostly stunted. The commonly observed ones are *Anogeissus latifolia* (Roxb.) Wall. ex Bedd., (*Bauhinia racemosa* Lam., *Chloroxylon swietenia* DC., *Cochlospermum religiosum* (L.) Alston, *Cassia fistula* L., *Careya arborea* Roxb., *Semecarpus anacardium* L.f., *Albizia lebbeck* Benth., *A. procera* Benth., *Ixora brachiata* Roxb. and *Radermachera xylocarpa* (Roxb.) K. Schum.

The climbers are very few and belong to *Hemidesmus indicus* R. Br., *Ventilago maderaspatana* Gaertn., *Similax zeylanica* L. and *Argyreia* spp.

The herbaceous flora is only seasonal and are contributed by grasses like *Apluda varia* Hack., *Aristida* sp., *Eragrostis unioloides* Nes, *Heteropogon contortus* P. Beauv., *Setaria glauca* P. Beauv., and other herbs like *Polycarpaea aurea* Wt. & Arn., *Crotalaria* spp., *indigofera* spp. and *Barleria buxifolia* L.

(ii) The dry deciduous forests

The dry deciduous hill forests are found on the eastern side of the Western Ghats at elevations between 500-1000 m. The rainfall varies between 800-2000 mm. The species composition and density of the species varies very considerably from place to place. Typically, the following species are chiefly present. *Diospyros montana* Roxb., *D. sylvatica* Roxb., *Eriolaena quinquelocularis* Wt., *sterculia urens* Roxb., *Anogeissus latifolia* (Roxb.) Wall. ex Bedd., *Butea monosperma* (Lam.) Taub., *Cassia fistula* L., *Emblica officinalis* Gaertn., *Grewia tilliaefolia* Vahl, *Pterocarpus marsupium* Roxb., *Terminalia* spp., *Albizia amara* (Roxb.) Boiv., *Bombax ceiba* L., *Mitragyna parvifolia* (Roxb.,) Korth., *Cochlospermum religiosum* (L.) Alston, *Givotia moluccana* (L.) Sreem., *Melia composita* Willd.

At comparatively higher rainfall areas the forests tend to become mixed deciduous type with a few evergreen elements. Most of the trees shed their foliage during December-February and remain in this condition till March-April, when the trees come to bloom. *Dendrocalamus strictus* (Roxb.) Nees and *Bambusa arundinacea* Willd. are also common in these forests.

Surubs are very few and are usually seen in open places. The common species are *Flacourtia indica* (Burm. f.) Merr., *Securinega leucopyrus* Brandis, *Carissa congesta* Wt., *Callicarpa tomentosa* Lam., *Colebrookea oppositifolia* Sm., *Xeromphis spinosa* (Thunb.) Keay, *Meyna laxiflora* Robins., *Ziziphus* spp. and the introduced *Lantana camara* L.

The common climbers are *Asparagus racemosus* Willd., *Cryptolepis buchanani* Roem. & Schult., *Calycopteris floribunda* Lam., *Cayratia pedata* (Lam.) Juss ex Gagnep., *Canavalia gladiata* DC., *Glycine wightii* (Wt. & Arn.) Verdc., *Dregea volubilis* Benth. and rarely *Entada pursaetha* DC.

(iii) Moist deciduous forests

The moist deciduous forests mostly occur between 500-100 m elevation at windward side where rainfall is comparatively high (2000-3000 mm). Many elements of dry deciduous forests, and at higher elevations evergreen elements also intrude into this zone. Important timber trees such as *Terminalia crenulata* Kurz, *Dalbergia latifolia* Roxb., *Lagerstroemia microcarpa* Wt., *Pterocarpus marsupium* Roxb., *Pterygota alata* R. Br., *Schleichera oleosa* (Lour.) Oken. *Tectona grandis* L.f. grow here luxuriantly. Other tree species of this zone are *Haldina cordifolia* (Roxb.) Ridsd., *Dillenia pentagyna* Roxb., *Miliusa tomentosa* (Roxb.) Sincl., *Terminalia alata* Heyne ex Roth, *T. paniculata* Roxb., *T. chebula* Retz., *Shorea roxburghii* G. Don, *Mitragyna parvifolia* (Roxb.) Korth., *Xylia xylocarpa* Antila & Ham., *Stereospermum colais* (Dillw.) Mabberly, *Vitex altissima* L., and *Bridelia retusa* Spreng. Although no distinct stratification can be observed, there are also a number of smaller trees of *Trema orientalis* Bl., *Kydia calycina* Roxb., *Clausena hestaphylla* Wt. & Arn., *Wrightia tinctoria* R.Br., *Litsea deccanensis* Gamble, *Grewia tiliaefoila* Vahl, *Cassia fistula* L., *Nothopodytes nimmoneana* (Grah.) Mabberly, *Olea dioica* Roxb., and *Oroxylum indicum* Vent. Another conspicuous feature of this zone is the luxuriant growth of *Bambusa arundinacea* Willd. which forms huge clumps in exposed places.

The moist deciduous forests almost resemble an evergreen forest during monsoon, when the trees form dense canopies with many large climbers overtopping the tree canop. These belong to *Gouania microcarpa* DC., *Ichnocarpus frutescens* R.Br., *Erythropalum populifolium* Mast., *Diploclisia glaucescens* (Bl.) Diels., *Dioscorea* spp., *Cayratia pedata* (Lam.) Juss. ex Gagnep., *Ventilago maderaspatana* Gaertn., *Teramnus labialis* Spreng., *Derris heyneana* Benth., *Naravelia zeylanica* DC., *Hiptage benghalensis* (L.) Kurz and *Mucuna* sp., A number of ferns such as *Drynaria quercifolia* (l.) Sm., *Lepisorus* spp. and orchids, particularly *Vanda* and *Dendrobium* sp., are also common on tree trunks in partially shaded localities.

The herbaceous flora is also very profuse in the cleared forest areas, particularly during rainy season. Some dominant ones in this category are *Cassia* spp., *Cleome viscosa* L., *Hibiscus lobatus* (Murr.) Kuntze., *Sida* spp., *Ageratum conyzoides* L., *Orassocephalum crepidioides* (Benth.) S. Moore, *Cynoglossum zeylanicum* (Burm. f.) R. Br., *Stachytarpheta* spp., *Rauvolfia serpentina* Benth. At certain other places *Eupatorium odoratum* L., is the only species constituting the ground cover.

Moist deciduous forest types normally occur on black soil in low rainfall areas and these show transition to semi-evergreen type (as evident by intrusion of evergreen elements), where higher rainfall occurs. Although no single species dominance can be observed in any given place, Aiyer (1932) recognized *Terminalia-Tectona-Haldina* association or *Terminalia-Grewia* type or *Anogeissus-Terminalia-Emblica* association in the moist deciduous forests of Western Ghats. There are also large patches of bamboo clumps, chiefly of *Bambusa arundinacea* Willd. and *Dendrocalamus hamiltonii* Munro in disturbed areas. Over exploitation and inadequate regeneration have greatly reduced the bamboo population in the area.

(iv) Semi-evergreen forests

The Semi-evergreen forests appear at elevations ranging from 500-1500 m and usually along the windward side. The rainfall is very heavy, 2000-3500 mm. This type forms an intermediate type between the evergreen forests and moist deciduous forests on hill slopes by the presence of characteristic species of both the zones. The dominant species of the upper storey are *Artocarpus hirsutus* Lam., *Euphoria longan* (Lour.) Steud., *Elaeocarpus tuberculatus* Roxb., *Hopea parviflora* Bedd., *Mangifera indica* L., *Sterculia guttata* Roxb., ex DC., *Terminalia paniculata* Roth, *Vateria indica* L., *Holoptelia integrifolia* Planch., *Tertrameles nudiflora* R.Br., *Vitex altissima* L., *Bischofia javanica* Bl. and *Xylia xylocarpa* Antila & Ham.

The second storey, where distinct, comprises of *Bischofia javanica* Bl., *Alstonia scholaris* R.Br., *Dillenia pentagyna* Roxb., *Hydnocarpus pentandra* Manilal *et al.*, Mallotus philippensis Muell.-Arg. and Syzygium sp. The undergrowth is constituted by numberous herbaceous species belonging to Acanthaceae, Lamiaceae, Leguminosae, Poaceae, Zingiberaceae, Asteraceae, etc.

Among the climbers mention can be made of *Combretum latifolium* Bl., *Gouania microcarpa* DC., *Hugonia mystax* L., *Ichnocarpus frutescens* R. Br., *Cansjera rheedii* Gmel., *Jasminum rottlerianum* Wall. ex DC. and *Pothos scandens* L.

(v) Evergreen forests

The evergreen forests are supposed to be the climax type of forests, also termed as 'wet evergreen forest', and montane subtropical evergreen forests'. In Western Ghats the evergreen forests have developed in areas receiving very heavy rainfall, 3500-7500 mm and elevations ranging from 500-2600 m or even more. Although there is a marked summer and winter season, the evergreen forests of Western Ghats at many places viz., Silent Valley show typical characters of Tropical Rain forests.

There is a bewildering wealth of floral diversity in these forests. The trees are quite luxuriant and reach enormous heights. Several trees exhibit distinct buttresses. Some such trees are *Hydnocarpus pentandra* Manilal *et. al., Fagraea ceilanica* Thunb., *Knema attenuata* (HK.f. & Th.) Warb. The forests at lower elevations exhibits distinct stratification. The top storey or the emergent layer (20-40 m) has trees with a straight bole and dark green crown. The dominant species of this tier are *Canarium strictum* Roxb., *Artocarpus hirsutus* Lam., *Diospyros ebenum* Koen., *Knema attenuata* (Hk. f. & Th.) Warb., *Neolitsea cassia* (L.) Kost., *Myristica dactyloides* Gaertn. *Garcinia* spp., *Palaquium ellipticum* (Salz.) Baill., *Vitex altissima* L., *Cullenia exarillata* Robyns., *Poeciloneuron indicum* Bedd., *Calophyllum polyanthum* Wall. ex Choisy, *Vateria indica* L. and a host of others which cannot be enumerated here. It may also be noted that all these species may not occur in any one given place and again their density/ dominance also vary from place to place. Species like *Palaquium ellipticum* (Salz.) Baill., *Cullenia exarillata* Robyns., *Myristica dactyloides* Gaertn., *Elaeocarpus glandulosus* Wall. ex Merr., and *Litsea floribunda* (Bl.) Gamble are dominant in Silent Valley forests; *Mesua ferrea* L., *Vitex altissima* L., *Aglaia elaeagnoidea* (A. Juss.) Benth., *Cassine glauca*(Rottb.) Kuntze, *Diospyros buxifolia* (Bl.) Hiern., are dominant in Annamalai and Palni hills *Calophyllum polyanthum* Wall. ex. Choisy, *C. apetalum* Willd., *Garcinia gummi-gutta* (L.) Rob., *Canarium strictum* Roxb., *Lophopetalum wightianum* Arn. are frequent in northern Western Ghats.

The endemic genus *Poeciloneuron* with two species, *P. indicum* Bedd. and *P. pauciflorum* Bedd. occur in the western Ghats only from Mysore southwards and conspicuously absent northwards. *Mesua ferrea* L.. also occurs from Mysore southwards. The Dipterocarpaceae are represented by *Dipterocarpus. Hopea, Shorea, Vatica* and *Vateria.*

Of these, *Dipterocarpus indicus* Bedd. occurs from Mysore southwards. *Hopea utilis* (Bedd.) Bole is endemic to Tinnevelly Hills. Similarly *Myristica malabarica* Lam. (Myristicaceae) and palms like *Bentinckia condapanna* Berry, *Pinanga dicksonii* Scheffer and about nine species of *Calamus* are restricted to southern Western Ghats.

The second storey trees are medium sized trees which are adapted to shady conditions. Some of these are *Litsea (floribunda)* (Bl.) Gamble, *Aporusa lindleyana* Baill., *Antidesma menasu* Miq., *Carallia brachiata* (Lour.) Merr., *Hydnocarpus laurifolia* (Dennst.) Sleumer, *Acrocarpus fraxinifolius* Wt., *Baccaurea courtallensis* Muell., (in southern Western Ghats), *Elaeocarpus glandulosus* Wall. ex Merr., *Holigarna ferruginea* March., *Persea macrantha* (Nees) Kosterm., *Pterospermum xylocarpum* (Gaertn.) Sant. & Wagh, *Cinnamomum* spp., *Holigarna beddomeii* Hk. f., *Macaranga tomentosa* Wt., *Psychotria* sp., *Sapindus laurifolius* Vahl,. *Meliosma simplicifolia* (Roxb.) Walp. and a host of others. As one traverses through these forests from northern Western Ghats towards southern region, the species composition varies considerably.

The ground flora is constituted by a number of shrubs and herbs mainly belonging to *Strobilanthes* spp., *Psychotria* spp., *Begonia malabarica* Lam., *Elatostema lineolatum* Wt., *Ophiorhiza murgos* L. *Impatiens* spp., *Girardinia diversifolia* (Link) Friis, *Scutellaria violaceae* Heyne ex Benth. and in certain places patches of Zingiberaceae members such as *Costus speciosus* (Koen.) Sm., *Globba* spp., *Catimbium malaccense* (Burm. f.) Holttum, *Curcuma decipiens* Dalz. and a number of Asteraceae members are common. Ground orchids like *Habenaria, Zeuxine, Disperis, Pecteilis gigantea* (Sm.) Rafin., *Nervillia aragoana* Gaud., *Malaxis rheedii* Sw. are also common. Along ravines and marshy areas the presence of huge patches of the prickly *Calamus* spp., *Angiopteris evecta* (Forst.) Hoffm., *Cyathea gigantea* (Wall. ex Hk.) Holttum with their graceous foliage is remarkable. In addition, root parasites like *Aeginetia indica* Roxb., *Balanophora* spp., and saprophytes like *Epipogium roseum* (D.Don) Lindl. are also common.

The epiphytic flora is also quite varied and maily belong to orchids such as species of *Dendrobium, Cymbidium, Eria, Bulbophyllum, Vanda, Aerides, Acampe, Oberonia, Pholidota* and ferns and fern-allies such as *Drynaria quercifolia* (L.) Sm., *Pyrrosia adnascens* (Sw.) Ching, and *Lycopodium* spp. The other flowering plants on these tree trunks are *Aeschynanthes perrottetii* A.DC. (with fleshy leaves and scarlet red flowers), *Hoya pendula* Wt., *Peperomia* spp., *Remusatia vivipara* Schott. and *Pothos scandens* L.

The climbers and lianas are also too many in these forests. *Adenia wightiana* (Wall. ex Wt. & Arn.) Engl., *Derris brevipes* Baker, *Mucuna hirsuta* Wt. & Arn., *Naravelia zeylanica* (L.) DC., *Sarcostigma keleinii* Wt. & Arn., *Thunbergia fragrans* Roxb., *Rubia cordifolia* L., *Entada* sp., *Ancistrocladus heyneanus* Wall. ex Grah., *Gnetum ula* Brongn., and *Allophylus serratus* (Roxb.) Kurz are a few important ones in this category.

The natural vegetation of the Western Ghats also includes sizeable areas of bamboo forest. 'Wet bamboo brakes' including *Oxytenanthera* spp. and *Bambusa kurzii* (Munro) Balakr. occur in dense clumps along streams in the evergreen and semi-evergreen forest belts.

The Sholas

The discussion on the vegetation types of Western Ghats would be incomplete without the mention of *'Sholas'* of the Nilgiris. The sholas are characteristically seen along the folds of rolling downs at a height of 1600 m and above, where moisture content is very high. They are isolated compact evergreen patches composed of stunted trees and bushes. Particularly at higher elevations the sholas exhibit an admixture of both tropical and temperate genera. These climatic climax forests are a highly threatened community today.

The species diversity is remarkably very high. The most conspicuous trees and shrubs in the sholas are *Hydnocarpus alpina Wt.*, *Michelia nilagirica* Zenk., *Berberis tinctoria Lesch.*, Mahonia *leschenaultii* (Wall. ex Wt. & Arn.) Takeda, *Garcinia gummi-gutta* (L.) Rob., *Gardenia obtusa* Wall. ex Wt. & Arn., *Ternstroemia gymnanthera* (Wt. & Arn.) Sprague, *lex denticulata* Wall. ex Wt., *I. wightiana* Wall. ex Wt., *Euonymus crenulatus* Wt., *Microtropis ramiflora* Wt., *Cinnamomum wightii* Meissn., *Meliosma wightii* Planch., *Pentapanax leschenaultii* Seem., *Schefflera ramosa* (Wt.) Harms., *Macaranga indica* Wt., *Acronychia pedunculata* (L.) Miq., *Atalantia wightii* Tanaka, *Eurya nitida* Korth., *Ligustrum robustum* (Roxb.) Bl., etc. The undergrowth consists mainly of *Clematis wightiana* Wall., *Polygala arillata* Ham., *Osbeckia leschenaultiana* DC., *Debregeasia longifolia* Wedd., *Maesa indica* Wall., and a few others. A rich growth of orchids, mainly of *Calanthe triplicata* (Will.) Ames. *Aerides ringens* (Lindl.) Fisch., *Habenaria longicornua* Lind., etc. are also observed in the epiphytic flora. The insectivorous plants like *Drosera* spp. and *Utricularia* spp. are common in the open meadows above 2000 m. The Temperate elements like the *Rhododendron arboreum* Sm., *Mahonia* sp. are also frequent along the margin of the shola forests.

The Shola forests also occur in Anamalai and Palni hills but being situated at comparatively lower heights, the vegetation is purely of tropical evergreen type. A number of semi-parasites like *Dendrophthoe falcata* (L.f.) Etting, *Helixanthera wallichiana* (Schult.) Danser, *Macrosolen parasiticus* (L) Danser and *Viscum* spp. are also common on branches of trees.

The open meadow bordering the Sholas support a variety of colourful herbs and shrubs like *Anemone rivularis* Wall., *Ranunculus reniformis* Wall., ex Wt. & Arn., *Cardamine hirsuta* L., *Viola betonicifolia* Sm. ssp. *betonicifolia*, *Hypericum mysorense* Heyne, *Impatiens nilagirica* Fisch., *Parnasia chinensis Franch.*, *Rhodomyrtus tomentosa* Wt., *Osbeckia cupularis* D.Don, *Vernonia arborea* Buch.-Ham., *Erigeron* spp. and several Fabaceae and Gentianaceae members.

Another remarkable feature of the vegetation of the Nilgiris of Western Ghats is their close affinity with the flora of Khasi and Jaintia Hills of Meghalaya. Some of the trees and shrubs common to these widely separated regions include *Ternstroemia gymnanthera* (Wt. & Arn.) Spreng., *Hypericum hookerianum* Wt. & Arn., *Thalictrum javanicum* Bl., *Turpinia nepalensis* Wall. ex Wt. & Arn., *Cotoneaster buxifolia* Wall., *Parnassia wightiana* Wall. ex Wt. & Arn., *Pentapanax leschenaultii* Seem, *Lonicera ligustrina* Wall., *Lactuca hastata* DC., *Gaultheria fragrantissima* Wall., *Rhododendron arboreum* Sm., *Lysimachia obovata* Ham. and *Symplocos laurina* Wall. ex G. Don. Fyson (1932) estimated that about 17 per cent of the flora of South Indian hill stations occurs in Khasi hills of Meghalaya and 12 per cent in temperate Himalaya.

Grasslands

Grasslands usually occur at higher elevations, particularly in southern part of Western Ghats. The trees and shrubs scattered here are all stunted. These grasslands are not climax vegetation but are maintained by seasonal burning. Characteristic grass species are

Saccharum spontaneum L., *Impertata cylindrica* (L.) Raeus., *Arundinella setosa* Trin., *Chrysopogon hackeilii* Fisch., *Eulalia trispicata* Kuntze, *Themeda triandra* Forssk. and *Jansenella griffithiana* (C. Muell.) Bor. A number of other angiosperm species also grow in these grasslands and these are *Hypericum japonicum* Thunb., *Osbeckia leschenaultiana* DC., *Vernonia* spp., *Anaphalis aristata* DC., *Strobilanthes kunthianus* T. Anders., *Plantago major* L., *Rumex nepalensis* Spreng., *Exacum bicolor* Roxb., *Lobelia nicotinaefolia* Roth ex Roem. & Schult. and several members of Acanthaceae and Lamiaceae. Most of the grasses dry up during summer when the entire grasslands are burnt but appear again with the onset of monsoon. *Lilium neilgherense* Wt., *Lobelia leschenaultiana* (Presl) Skottsb. in the Nilgiri region are quite interesting. On the hill slopes *Phoenix humilis* Becc. & Hk. f. is another common species of the grasslands. *Emblica officinalis* Gaertn. also makes its scattered appearance in these grasslands.

Plant Diversity

The Western Ghats phyto sociological region is another major genetic estate with an enormous biodiversity of ancient lineage. As many as 400 species of flowering plants occur here of which 56 genera and 2100 species are endemic (Nayar, 1982). Poaceae (120 genera and 400 species), Leguminosae (85 genera and 220 species), Orchidaceae (60 genera and 250 species), Acanthaceae (55 genera and 165 species), Cyperaceae (21 genera and 160 species), Euphorbiaceae (55 genera and 140 species) are a few dominant families in the Western Ghats flora. Apart from the valuable timber species such as the species of *Calophyllum, Garcinea, Mamaea* of the family Clusiaceae, *Dipterocarpus, Hopea, Shorea, Vatica, Vateria* of Dipterocarpaceae, *Pterocarpus marsupium* Roxb., *Xylia xylocarpa* (Roxb.) Taub., *Albizia* spp., Dalbergia spp., *Acrocarpus fraxinifolius* Wt. & Arn., *Bauhinia malabarica* Roxb., of Leguminosae, *Cullenia exarilata* Robyns. (Bombacaceae) and several Lauraceae members, bamboos are also well represented here. There are about 8 genera and 24 species of bamboos.

The Western Ghats are also a rich germplasm centre for a number of wild relatives of economically important species of **Cereals and Millets :** *Panicum psilopodium* Trin., *Oryza coaractata*

Roxb., *Pennisetum glaucum* (L.) R. Br., *Chionachne koenigii* Thw., *C. semiteres* Henr., *Coix gigantea* Roxb., *Trilobachne cookei* Henr.

Legumes : The Western Ghats are a major centre of diversity for the following legume species. *Atylosia albicans* (Wt. & Arn.) Benth., *A. goensis* Dalz., *A. trinervia* (DC.) Gamble, *A. elongata* Benth., *A. platycarpa* Benth., *A. graniflora* Benth., ex Baker, *A. kulnensis* Dalz., *A. lineata* Wt. & Arn., *A. mollis* Benth., *A. nivea* Benth., Wt. & Arn., *A. scarabaeoides* (L.) Benth., *A. sericea* Benth. ex Baker, *A. villosa* Benth. ex Baker, *Canavalia virosa* (Roxb.) Wt. & Arn., *C. maritima* (Aubl.) Thouars, *Macrotyloma uniflorum* (Lam.) Verdc., *Sphenostylis bracteata* (Baker) Gillet, *Mucuna pruriens* (L.) DC., *Vigna sublobata* (Roxb.) Babu & Sharma, *V. vexillata* (L.) A. Rich. *V. pilosa* (Willd.) Baker, *V. umbellata* (Thunb.) Ohwi & Ohash., *V. dalzelliana* (Kuntze) Verdc., *V. mungo* (L.) Hepper, *V. dalzelliana* (L.) Wilczek, *V. khandalensis* (Sant.) Raghavan & Wadhwa.

Tropical and subtropical fruits : The Western Ghats harbour a large number of tropical fruits like *Artocarpus heterophyllus* Lam., *A. lacucha* Roxb., *Garcinia indica* Choisy, *Diospyros* spp., *Ensete superbum* (Roxb.) Cheesman, *Mangifera indica* L., *Mimusops elengii* L., *Spondias pinnata* (L.f.) Kurz, *Vitis* spp., *Ziziphus oenoplia* (L.) Mill., *Z. rugosa* Lam., *Rubus ellipticus* Sm., *R. niveus* Thunb., *R. alceifolius* Poir.

Vegetables : *Abelmoschus angulosus* Wall. ex Wt. & Arn., *A. moschatus* Medic., *A. manihot* L.,. *A. ficulneus* (L) Wt. & Arn., *Amorphophallus paeonifolius* (Dennst.) Nicolson, *Cucumis setosus* Cogn., *C. callosus* (Rottl.) Cogn., *Luffa gravelones* Roxb., *Momordica cochinchinensis* (Lour.) Spreng., *M. subangulata* Bl., *Solanum indicum* L., *Trichosanthes anamalaiensis* Bedd., *T. bracteata* (Lam.) Voigt, *T. cusidata* Lam., *T.perrottetiana* Cogn., *T. neriifolia* L., *T. villosula* Cogn.

Oilseed types : *Sesamum laciniatum* Klein and *S. prostratum* Retz.

Spices and condiments : *Cinnamomum zeylanicum* Bl., *Myrstica dactyloides* Gaertn., *M. malabarica* Lam., *Piper nigrum* L. (black pepper), *P. schmidtii* Hk. f., *P. longum* L. (White pepper), *Zingiber purpureum* Rosc., *S. officinale* Rosc., *Z. zerumbet* Sm. and *Elettaria cardamomum* Maton (cardamom). In addition, wild relatives of coffee and sugarcane are also well represented here.

The diversity in medicinal plants is also to high in this zone, but only a few important ones can be mentioned : *Acacia sinuata* (Lour.) Merr. (Mimosaceae), *Acalypha racemosa* Wall. ex Benth. (Euphorbiaceae), *Acorus calamus* L. (Araceae), *Adhatoda zeylanica* Medic. (Acanthaceae), *Alangium salviifolium* (L.f.) Wang. (*Alangiaceae*), *Alstonia scholaris* R. Br. (Apocynaceae), *Anamirta cocculus* Wt. & Arn. (*Menispermaceae*), *Aristolochia indica* L. (Aristolochiaceae), *Artemisia nilagirica* (Cl.) Pamp., (Asteraceae), *Bauhinia racemosa* Vahl (*Caesalpiniaceaee*). *Buddleja asiatica* Lour. (Budlejaceae), *Canarium strictum* Roxb., (Burseraceae), *Cardiospermum halicacabum* L. (Sapindaceae), *Cassia* spp. (Caesalpiniaceae), *Ceropegia* sp. (Asclepiadaceae), *Dioscorea* spp. (Dioscoreaceae), *Costus speciosus* Sm. (Zingiberaceae), *Entada pursaetha* DC. (Mimosaceae), *Ficus exasperata* Vahl (Moraceae), *Gloriosa superba* . (Liliaceae), *Gmelina arborea* Roxb. (Verbenaceae), *Gymnema sylvestre* (Retz.) R. Br. ex Schult. (Asclepiadaceae), *Helicteres isora* L. (Sterculiaceae), *Holarrhena antidysenterica* (Roth) A.DC. (Apocynaceae), *Ichnocarpus frutescens* (L.) R. Br. (Apocynaceae), *Iphigenia indica* A. Gray (Liliaceae), *Knema attenuata* (Hk. f & Th.) Warb. (Myristicaceae), *Leucas* spp. (Lamiaceae), *Mallotus philippensis* (Lam.) Muell. - Arg. (Euphorbiaceae), *Mesua nagassarium* (Burm. f.) Kost (Clusiaceae), *Momordica* spp. (Cucurbitacae)., *Murraya paniculata* (L.) Jack. (Rutaceae), *Myristica dactyloides* Gaertn., (Myristicaceae), *Naregamia alata* Wt. & Arn. (Meliaceae), *Pterocarpus marsupium* Roxb. (Fabaceae), *Rauvolfia* spp. (Apocynaceae), *Rourea minor* (Gaertn.) Alston (Connaraceae), *Scilla hyacinthina* Mc. Br. (Liliaceae), *Smilax zeylanica* L. (Smilacaceae), *Solanum* spp. (Solanaceae), *Stephania japonica* Miers. (Menispermaceae), *Strychnos* spp. (Loganiaceae), *Terminalia* spp. (Combretaceae), *Tinospora cordifolia* (Willd.) Miers. (Menispermaceae), *Trema orientalis* Bl. (Ulmaceae), *Tylophora* sp., *Dregea volubilis* (L.f.) Benth. ex Hk. f. (Asclepiadaceae) and *Wrightia tinctoria* (Roxb.) R. Br. (Apocynaceae).

The life support species which offer very valuable subsidiary food are *Alangium salviifolium* (L.f.) Wang. (Alangiaceae), *Antidesma acidum* Retz., *A. menasu* Miq. ex Tul. (Euphorbiaceae), *Artocarpus* spp. (Moraceae), *Baccaurea courtallensis* Muell. - Arg. (Euphorbiaceae), *Calamus rotang* L. (Arecaceae), *Canthium travancoricum* Bedd. (Rubiaceae), *Emblica officinalis* Gaertn. (Euphorbiaceae), *Ficus* spp. (Moraceae), *Flocourtia indica* (Burm. f.) Merr. (Flacourtiaceae),

Mangifera indica L. (Anacardiaceae), *Phoenix humilis* Becc. & Hk. f. (Arecaceae), *Physalis* spp. (Solanaceae), *Rubus* spp. (Rosaceae), *Solanum* spp. (Solanaceae), *Syzygium cumini* (L.) Skeels (Myrtaceae), *Caryota urens.* L. (Arecaceae) and several *Colocasia* and *Alocasia* spp. (Araceae).

Chapter 11

Phyto Sociological Region of the Deccan Peninsula

The Deccan Peninsular phyto sociological region in India is the most extensive but relatively a homogeneous zone covering about 43 per cent of the total Indian landmass. Major portion of Maharashtra, Madhya Pradesh, Karnataka, Tamil Nadu, Andhra Pradesh, Orissa and Bihar are covered by this zone. 'Eastern Ghats'

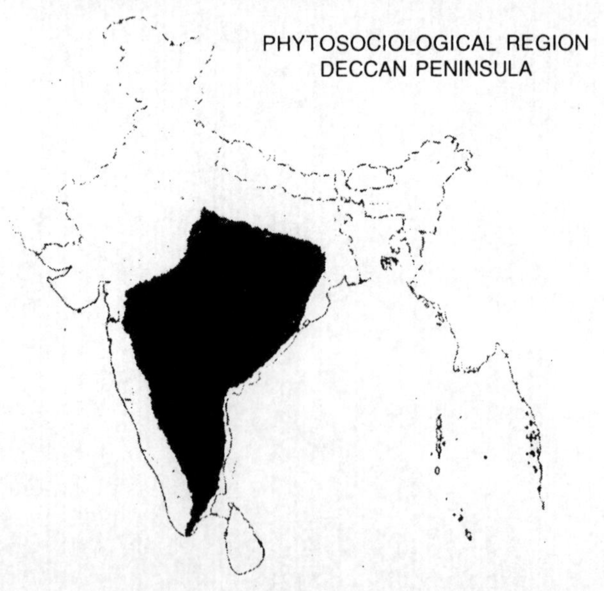

of some phytogeographers, which is a narrow mountainous belt extending from Nilgiri hills in the south-west to Gurjahat hills in the north-east over a distance of 800 km is also merged under the present biogeographic zone. Mainly the zone is covered by deciduous forests, thorn forests and scrublands. Evergreen forests are confined to certain pockets in Eastern Ghats. The deciduous forests are dominated by Sal and Teak and the thorn forests by *Acacia-Albizia* association. Although the floristic diversity is comparatively much lower, this zone has several endemic species of great botanical value.

The region reveals a considerable variation in climate. The summer is long and very severe. The average mean daily maximum temperature in the northern part goes up to 39°C and the average daily minimum is 15°C. February to September are the hotter months with May being the hottest month of the year. The winter is very mild and confined to December-January.

The rainfall pattern is also highly variable. The mean annual rainfall ranges from just 70 mm (in northern plains of Karnataka, Maharashtra) to 3000 mm in parts of Eastern Ghats. The rainfall is received mostly from S.W. monsoon during June to August and during August to September from N.E. monsoon.

Soil types vary from black cotton soil in the western part (Maharashtra, Karnataka, Madhya Pradesh) to red soil in the south and north-east part of the Peninsula (Andhra Pradesh, Tamil Nadu, Bihar, Orissa).

The average elevation is about 600 m, though there are peaks as high as 2637 m.

Vegetation

The flora and vegetation of the Deccan region is fairly well known through the several publications (Cooke, 1901-08; Haines, 1915-20; Gamble, 1915-36; Rao and Razi, 1974; Oommachan, 1977; Sharma *et. al.*, 1984; Nair and Henry, 1983; Saldanha, 1984). The vegetation can be classified mainly under the following types viz. (1) Dry deciduous type, (2) Thorn moist deciduous type, (3) Thorn scrub forests and (4) Semi-evergreen type.

1. *Dry Deciduous Forests*

The dry deciduous forests occupy bulk of the area. Most of the deciduous species noticed in other biogeographic zones are also recorded here. Some of common species in this type of forest are *Boswellia serrata* Roxb. ex Coleb., *Lannea coromandelica* (Houtt.) Merr., *Sterculia urens* Roxb., *Lagerstroemia parviflora* Roxb., *Butea monosperma* (Lam.) Taub., *Diospyros melanoxylon* Roxb., *Anogeissus latifolia* (Roxb.) Wall. ex Bedd., *Acacia catechu* (Roxb.) Willd., *A. torta* (Roxb.) Craib, *Emblica officinalisa* Gaertn. *Cassine glauca* Lam., *Ziziphus mauritiana* Ham., *Nyctanthes arbor-tristis* L., *Gardenia turgida* Roxb., *Ehretia laevis* Roxb., *Cochlospermum religiosum* (L.) *Alston* and *Albizia* spp. *Dendrocalamus strictus* Nees also occurs in moist localities.

The shrub layer is very poor but in open places shrubs like *Holarrhena antidysenterica* (Roth) A. ḌC., *Grewia hisuta* Vahl, *Flacourtia indica* (Burm. f.) Merr., *Capparis* spp., *Kirganelia reticulata* Baill., *Securinega virosa* Baill., *Casearia elliptica* Willd., *Woodfordia fruticosa* Kurz and lxora spp. are common.

These shrubs and trees are often covered by twiners and stragglers like *Smilax zeylanica* L., *Asparagus racemosus* Willd., *Ampelocissus* spp., *Hemidesmus indicus* R.Br., *Ichnocarpus frutescens* R.Br., *Cryptolepis buchanani* Roem. & *Schult*., Atylosia scarabaeoides Benth. and a few others. Among the herbaceous flora maily grasses legumes and few Lamiaceae members are conspicuous.

2. *Moist Deciduous Forests*

The moist deciduous forests occur on the sheltered lower slopes of hills where moisture content is high. The species composition of this forest slightly varies with that of dry deciduous ones. The forest has tall trees of *Shorea robusta* Gaertn. (in Madhya Pradesh), *Terminalia* spp., *Pterocarpus marsupium* Roxb., *Dalbergia paniculata* Roxb., *Haldina cordifolia* (Roxb.) Ridsd., *Schleichera oleosa* (Lour.) Oken., *Garuga pinnata* Roxb., *Mallotus philippensis* Muell. - Arg., *Anogeissus latifolia* (Roxb.) Wall. ex Bedd., *Cassia fistula* L., *Buchanania lanzan* Spreng., *Soymida febrifuga* Juss. and occasionally *hardwickia* binata *Roxb.* At several places the understorey is not well defined. The ground cover is also very poor. Grasses are found to occur either as an undergrowth or as a pure crop in open clay soils.

A good development of Teak (*Tectona grandis L.f.*) associated with *Anogeissu latifolia* (Roxb.) Wall. ex Bedd., *Terminalia crenulata* Kurz, *Lagerstroemia parviflora* Roxb. and *Diospyos melanoxylon* Roxb. can also be observed in several places. In the northern and central part of the zone, *Sal* is typically associated with *Terminalia* spp., *Pterocarpus marsupium* Roxb. and *Dalbergia paniculata* Roxb.

Sometimes it may also be noted that *Anogeissus latifolia* (Roxb.) Wall. ex Bedd., *Butea monosperma* (Lam.) Taub., *Acacia catechu* Willd. and *Terminalia alata* Roth become extremely dominant so as to form subtypes of pure stands. This type of vegetational mosaic has been correlated with the edaphic factors. *Terminalia crenulata* Kurz is predominant in Madhya Pradesh where the soil is moist or water logged. Similarly *Chloroxylon swietenia* DC. is the main species on the rocks of Vindhyan formation, where the soil is shallow, arid and sandy. *Acacia leucophloea* Willd. occurs in pure stands on black cotton or poor shallow soils. In the Nullamalai region of Eastern Ghats there are some characteristics localised types of vegetation, each one named after the dominant species in the vegetation such as *Boswellia* type, *Terminalia* type, *Phoenix* type and *Calamus* type.

3. Thorn Scrub Forests

The thorn scrub forests appear in much drier areas as in the northern interior Karnataka and southern parts of Maharashtra and in parts of Tamil Nadu. Here, the summer temperature goes up 40°-45°C and annual rainfall is too meagre with only 50-60 mm. The long dry season of 7-8 months has led to the development of thorn scrub forest over a large tract in this biogeographic zone. The vegetation is composed of *Carissa congesta* Wt., *Canthium parviflorum* Lam., *Cassia auriculata* L., *Dodonaea viscosa* Jacq. *Naringi alata* Wt. & Arn., *Erythroxylum monogynum* Roxb., *Pterolobium hexapetalum* (Roth) Sant. & Wagh, *Rhus mysorensis* Heyne, *Parkinsonia aculeata* L., *Woodfordia fruticosa* Kurz, *Cadaba fruticosa* (L.) Druce, *Jatropha glandulifera* Roxb., *Microcos paniculata* L., *Grewia abutilifolia* Juss., *Flacourtia indica* (Burm. f.) Merr., *Ixora brachiata* Roxb., *Acacia chundra* Willd., etc. There are also species like *Caralluma umbellata* Haw., *Coleus canisus* (Roth) Vatke, *Sarcostemma acidum* (Roxb.) Voigt, etc. which are all adapted to the increasing dryness of the region. Under the thorny thickets where there is shade and accumulation of humus, herbaceous species of *Polygala*, *Mollugo*, *Biophytum*,

Dicliptera, Phaulopsis, etc. come up. But in open gravelly soil, herbs like *Andrographis serpayllifolia* Wt. *Stylosanthes fruticosa* (Retz.) Alston *Portulacca* spp. are seen among grasses.

4. *Semi-Evergreen Forests*

In the Deccan Peninsular region the semi-evegreen forests are confined to certain limited pockets above 500 m in the Eastern Ghats. The annual rainfall goes up to 500 cm per year. Some of the important species here are *Sterculia guttata* Roxb., *Dysoxylum binectariferum* Hk. f., *Pittosporum nepaulense* (DC.) Rehd. & Wils., *Euphoria longan* (Lour.) Steud., *Wendlandia thyrsoidea* (Roem. & Schult.) Steud., etc. In the Nullamalai region of Andhra Pradesh a sort of dry evergreen forest composed of *Canthium dicoccum* (Gaertn.) Merr., Manilkara *Hexandra* (Roxb.) Dub., *Memecylon umbellatum* Benth., Santalum album L., Stryhnos nux-vomica L., etc. has developed.

Diversity

The Deccan phyto sociological region, particularly the Eastern Ghats region shows a remarkable diversity in flora. As many as 2500 species of vascular plant species are expected here, of which about 4 per cent are endemics. There are many species of wild relatives of our cultivated plants. *Oryza meyeriana* (More & Steud.) Baill., the mountain rice appearing like seedling of bamboo in northern Eastern Ghats, *O. jeyporensis* Govind. & Krishn. an endemic rice of wetlands in Eastern Ghats, *O. officinalis* Wall. ex Watt ssp. *Malampuzhaensis* (Krishn. & Chandras.) Tateoka in drier tracts of Nullamalais, the wild banana, *Musa balbisiana* Colla, in Madgole hills and the numerous medicinal plants, wild food plants, etc. are only a few to mention. Shifting cultivation locally known as "podu", excessive grazing and over exploitation of forests for fuel wood in the eco-sensitive zone have depleted the forest cover considerably rendering a number of plant species endangered.

Chapter 12
Phyto Sociological Region of the Indian Coasts

The coast line of India stretches from Gujarat to Cape Comorin (Kanyakumari) in the west and from Sunderbans to Cape Comorin in the East. Approximately 5400 km long stretch of coast line in the mainland has very diverse set of biotic communities. In addition, the Andaman and Nicobar group of Islánds also contain some of the least disturbed and best preserved coastal vegetation.

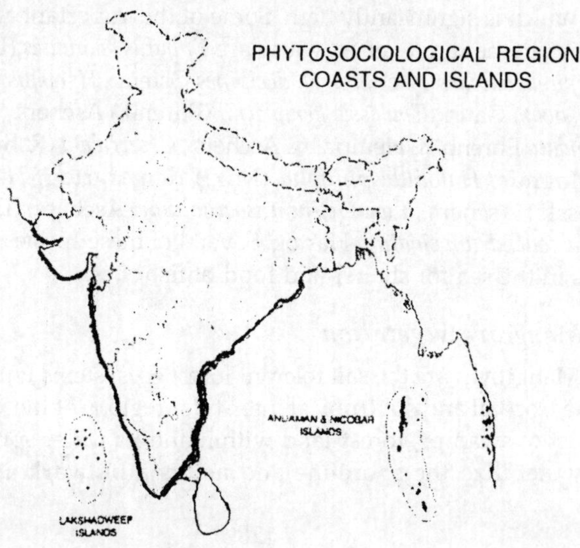

The coastal ecosystems of India include the (a) submerged vegetation, (b) mangrove forests (c) strand vegetation and (d) the tidal or swamp forests. The latter two can be broadly grouped as beach forests.

(a) Submerged vegetation

Submerged vegetation of the Indian coasts includes the marine algae and the sea grasses. Marine algae are found in sandy beaches with shallow water stream with rocks and boulders. Also the coral reefs in the sea facilitate a luxuriant growth of marine algae. Diverse genera belonging to green algae (Chlorophyceae), brown algae (Phaeophyceae) and red algae (Rhodophyceae) are specially dominant. The main genera are *Enteromorpha, Caulerpa, Codidum, Ulva, Halimeda, Dictyosphaeria* (Chlorophyceae); *Padina, Turbinaria, Sargassum* and *Dictyota* (Phaeophyceae); and *Porphyra, Halymenia, Gracilaria* (Rhodophyceae).

The sea grasses (Flowering plants) are common on sandy soils in quiet shallow waters beneath the ebb level. They are an important group of primary producers of marine and brack water ecosystem. They absorb and circulate the minerals which are in the sediment form. An interesting study conducted by Quasim and Bhattachari (1971) reveals that the maximum daily productivity of a mixed community of *Thalassia hemprichii* (Ehrenb.) Aschers. and *Cymadocea isoetifolia* Aschers. to be 14 gm of organic matter/m^2/ day which is significantly high. Some of the important angiosperm species in the Indian coast water are *Enhalus acoroides* (L.f.) Royle, *Halophyla baccarii* Aschers., *H. decipines* Ostenf., *H. ovalis* (R.Br.) Hk. f., *H. ovata* Gaud., *Thalassia hemprichii* (Ehrenb.) Aschers., *Cymadocea rotundata* Ehrenb. & Hempr. ex. Aschers., *C. serrulata* (R. Br.) Aschers. & Magnus, *Halodule pinifolia* (Miki) den Hartog, *H. uninervis* (Frossk.) Aschers., *Syringodium isoetifolium* (Aschers). Dandy and *Thalassodendrum ciliatum* Hartog. Several of these species form ideal beds in the sea for shelter and food of fishes.

(b) Mangrove Vegetation

Mangroves are the salt tolerant forest ecosystems found mainly in the tropical and subtropical inter-tidal regions of the world. The consist of swamps, forest-land within and its water spread areas. They stabilize the shoreline and act as a bulwark against the encroachment by the sea and the consequent inevitable erosion of the soil in the embankments.

The Indo-pacific region is known for its luxuriant mangroves. The total area of mangroves in India is estimated to be 6740 km^2 which is about 7 per cent of the world's mangroves. Out of these, the Sunderbans has the largest are under mangroves (4200 km^2); the next being the Andaman and Nicobar Islands (1190 km^2), both together accounting for 80 per cent of the mangroves in the country. The remaining are scattered in coasts of Andhra Pradesh, Tamil Nadu, Orissa, Maharashtra, Gujarat, Goa and Karnataka. It is of interest of observe that India has mangroves of the deltaic estuarine, brackwater and sheltered, insular bay types.

The perennial supply of fresh water along the deltaic coast and the admixture with the sea water through tides has resulted in the species richness. The diversity of species composition is evolved as a result of varying degree of adaptation to salt water-brackish water mix by the species.

The mangroves are physiological halophytes. The habitat of the mangroves impregnated with salt water is a physiologically dry one. The salt in water impedes absorption. The leaves possess halophilous properties, such as thick cuticle, large mucilage cells, etc. The buttress, knee, stilt roots, vertical pneumatophores are other adaptations. The physiological requirements of the plant species to withstand equally silt laden flood water, fresh water, euryhaline estuarine water, tolerance to sea water, etc., determine the composition and distribution of the mangroves.

The classification of the Indian mangroves has been attempted by several workers like Prain (1903), Blatter (1905), Cooke (1908), Curtis (1933), Navalkar (1952), Qureshi (1957), Champion and Seth (1968). This is primarily based on the tidal flow, soil salinity and nature of the communities. The species composition and dominance in different mangrove ecosystems also vary considerably. Some of the dominant mangrove species are *Rhizophora mucronata* Lam., *R. apiculata* Bl., R *Stylosa* Griff., *Bruguiera gymnorrhiza* (L.) Lam., *B. parviflora* (Roxb.) *Wt. & Arn.* ex Griff., *Ceriops tagal* (Perr.) C.B. Rob., *Avicennia marina* (Forssk.) Vierh., *A. officinalis* L., *Lumnitzera littorea* (Jack.) Voigt, *L. racemosa* Willd., *Aegiceras corniculatus* Blanc., *Xylocarpus granatum* Koen., *Sonneratia* spp. and rare (but common in Andaman and Nicobar Islands) mangrove palm *Nypa fruticans* Wurmb with short stem and large leaves arising from the ground level.

Table 12.1 : Area-wise Distribution of Mangroves

	State/Union Territory	*Area in Sq. km. (Values rounded to the nearest 10 sq. km.)*
1.	Andaman and Nicobar groups (including Nicobar islands)	1190
2.	West Bengal (the Sunderbans)	4200
3.	Orissa (all deltaic and coastal mangroves including the Mahanadi area)	150
4.	Andhra Pradesh (the Godavari and the Krishna)	200
5.	Tamil Nadu (the Cauvery and adjacent coastal stretch)	150
6.	Karnataka (Coondapur and Malpe, Karwar Patches)	60
7.	Goa	200
8.	Maharashtra (inclusive of Ratnagiri, Vijayadurg, Malvan, Devgad, etc. inclusive of 85 sq. km)	330
9.	Gujarat (Including the Narmada, the Tapati Gulf of Kambhayat mangroves of 105 sq. km.)	260
	Total	**6740**

Source : Anonymous, 1987 a.

The shrubs mainly comprise *Aegiatilis rotundifolia* Roxb., and *Acanthus iliciolius* L. (with spiny margined leaves and blue flowers) on poor saline plains. The mangrove fern *Acrostichum aureum* L. is also common in certain localities. The other herbaceous succulent halophytes are *Suaeda maritima* (L.) Dum., *Sesuvium portulacastrum* L., *Salicornia brachiata* Roxb. and iana several places *Aegiceras cornaiculatus* Blanc. In the mixed zone of sea and fresh water, the salt tolerant fresh water plants *Bruguiera gymnorrhiza* (L.) Lam., *B. pauciflora* (Roxb.) Wt. & Arn. ex Griff., *Excoecaria agallocha* L., *Sonneratia apetala* Buch. Ham., *Heritiera formes* Buch.-Ham., *Lumnitzera racemosa* Willd. and *Dalbergia candenatensis* (Dennst.) Prain, are seen. *Nypa fruticans* Wurmb, *Phoenix paludosa* Roxb. and *Aegiatilis rotundifolia* Roxb. have been recorded in shallow waters. The last is reported only from Sunderbans and Mahanadi delta. The 'Sundri' trees (*Heritiera minor* Lam.) have become very rare and *Kandelia candel* (L.) Druce has totally disappeared in several localities.

Mangroves of Mahanadi Delta in Orissa

The common mangroves here are *Rhizophora mucronata* Lam., *R. apiculata* Bl., *Bruguiera gymnorrhiza* (L.) Savi, B. *cylindrica* (L.) Bl., *B. sexangula* (Lour.) Poir., *Ceriops decandra* (Griff.) Ding Hou and *Excoecaria agallocha* L. Species such as *Kandelia candel* (L.) Druce, *Rhizophora apiculata* Bl., *Phoenix paludosa* Roxb. have disappeared if not extinct.

The sea shores are dominated by herbaceous and shrubby species like *Acanthus ilicifolius* L., *Salicorniaa bractiata* Roxb., *Suaeda maritima* (L.) Dum., and *S. monoica* Forssk. and these species mark the transitional stage from saline blank areas to trees or shrub covered forest areas.

Mangroves of Andaman and Nicobar Islands

The Andaman and Nicobar Islands contain some of the least disturbed and best preserved mangroves. The floristic diversity is very low and the forests are largely composed of a single species or of a few allied ones. The mangrove forests are well developed along the mud flats, estuaries, borders of lagoons and creeks. The dominant mangrove species are *Rhizophora mucronata* Poil., *R. stylosa* Griff. (on seaward fringe of swamp where the water salinity is very high) and *Bruguiera gymnorrhiza (L.)* Savi. The other common ones are *Avicennia marina* (Forssk.) Vierh., *Xylocarpus granatum* Koen., *Lumnitzera littorea* (Jack.) Voigt, *L. racemosa* Willd., *Aegiceras corniculatum* (L.) Blanc., *Sonneratia caseolaris* (L.) Engl., *Excoecaria agallocha* L. (near the mangrove swamps), *Nypa fruticans* Wurmb along the tidal creeks and *Phoenix paludosa* Roxb. along the swampy creeks. The preponderance of the woody species of the mangroves is noteworthy.

Mangrove vegetation of Sunderbans

Sunderbans are formed in the vast delta complex of the river Ganga and river Brahmaputra. The tidal mangrove consists of *Ceriops decandra* (Griff.) Ding Hou, *Avicennia alba* Bl., *Rhizophora mucronata* Lam. *Xylocarpus granatum* Koen.

Mangroves of Godavari Delta in Andhra Pradesh

Pure crops of *Avicennia marina* (Forssk.) Vierh., *A. officinalis* L., A. *alba* Bl., impressive grassy pastures, extensive mudflata banks are common here. The muddy shores have vegetation chiefly of *Rhizophora mucronata* Lam., *R. apiculata* Bl., *Ceriops decandra* (Griff.) Ding Hou, *Bruguiera gymnorrhiza* (L.) Lam., and *Acanthus illicifolius* L. On tidal marshes *Lumnitzera racemosa* Willd. and *Excoecaria agallocha* L. are dominant and on creeks *Avicennia officinalis* L. is common.

The mangrove vegetation of the coastal Tamil Nadu has more or less the same species composition as that of Andhra Pradesh.

The Mangroves of West Coast of India

The West coast of India comprises about 12 per cent of the total mangroves area. *Ca* 41 species belonging to 30 genera and 21 families are listed as mangrove species. Some of the common species in this zone are *Ceriops tagal* (Per.) C.B. Rob., *Salvadora persica* L, *S. oleioides* Decne., *Avicennia officinalis* L., *Sonneratia apetala* Buch.-Ham., *Rhizophora mucronata* Lam. and the halophytes like *Suaeda maritim* (L.) Dum., *Salicornia brachiata* Roxb., *Atriplex stocksii* Boiss. and *Sesuvium portulacastrum* L.

The mangroves of Goa have been extensively studied by Untawale *et al.*, (1980, 82). The dominant species here are *Rhizophora maucronata* Lam., *Sonneratia alba* Sm., and *Avicennia officinalis* L. The other co-dominant species are *Rhizophora apiculata* Bl., *Sonneratia caseolaris* (L.) Engl., *Kandelia candel* (L.) Druce, *Bruguiera gymnorhiza* (L.) Lam., *Aegiceras corniculata* Blanc., *Derris heptaphylla* (L.) Merr., and *Excoecaria agallocha* L. It is interesting to note that *Kandelia candel* (L.) Druce and *Sonneratia caseolaris* (L.) Engl., which had a wider distribution in the past are now restricted to the Goa region.

III. Beach Vegetation

The Beach vegetation starts from the proximity of sea shore to about 30-50 m interior depending upon the local conditions. The area are not usually inundated by sea water except duing storm when high waves splash over them.

The Beach vegetation includes the open herbaceous dune formations and wooded beach forests towards the interior. The important herbaceous elements include *Ipomoea pes-capre* (L.) Sw., a creeper with large pink flowers and bilobed leaves. The effective sand binder forms extensive pure patches or associated with *Cyperus arenarius* Retz., *Aleuropus lagopoides* (L.) *Trin.*, *Drimia indica* (Roxb.) Jessope, *Scilla hyacinthina* (Roth) Macbr., *Iphigenia indica* Gray, *Sesuvium portulacastrum* L., *Phyla nodiflora* (L.) Greene, *Apluda mutica* L., *Ischaemum* Indicum (Houtt.) Merr., *Spinifix littoreus* (Burm. f.) Merr. Shrubs like *Scaevola sericea* Vahl (in Andaman and Nicobar Islands), *Dodonaea* Viscosa L., *Clerodendrum inerme* Gaertn., *Derris trifoliata* Lour., *Desmodium umbellatum* DC., *Pluchea tomentosa* DC., *P. indica* (L.) Juss., *Breynia* spp., are also common and associated with such climbers as *Asparagus racemous* Willd., *Cyclea peltata* Hk. f. & Th., *Cocculus hirsutus* (L.) Diels., *Caesalpinia bonduc* (L.) Roxb.

The forest vegetation which occurs behind the open dune formations includes mainly the ligneous species such as *Pongamia Pinnata* Merr., *Thespesia populnea* Soland., *Pithecellobium dulce* Benth., *Cerbera odollam* Gaertn., *Calophyllum inophyllum* L., *Hernandia peltata* Meissn., *Pandanus* spp., and in Andamans *Mimusops littoralis* Kurz, *Intsia bijuga* (Coleb.) Kuntze, *Syzygium samarangense* (Bl.) Merr. & Perry and palm-like *Cycas rumphii* Miq. (Gymnosperm) are dominant. *Hyphaene indica* Becc. is another remarkable palm which forms extensive patches on sandy beds of Goa region in the west coast.

In swampy places, species like *Barringtonia racemosa* Roxb., *Cynometra ramiflora* L., *Lumnitzera racemosa* Willd., *Sonnerati caeseolaris* (L.) Engl. and *Phoenix paludosa* Roxb. (particularly in Andamans) are common. Herbaceous ones include the ferns like *Acrostichum aureum* L. and the Acanthaceae member *Acanthus illicifolius* L.

The coastal vegetation at several places is being replaced by plams like *Borassus flabellifer* L., *Cocos nucifera* L. and *Casuarina equisetofolia* Forst., which act as a wind screen. In Andaman Islands the oil palm *Elaeis guyanensis* Jacq. is extensively planted and has now almost naturalized.

Chapter 13

Phyto Sociological Regions of Andaman and Nicobar Islands

The Andaman and Nicobar Islands are an elongated north-south oriented group of 348 Islands in the Bay of Bengal stretching for 590 km from 6°45′ - 13°4′ N and 92°12′ - 93°57′ E. The Andaman Islands are about 190 km distant from Cape Negrais in Burma, the nearest point in the mainland. Five Islands close together constitute

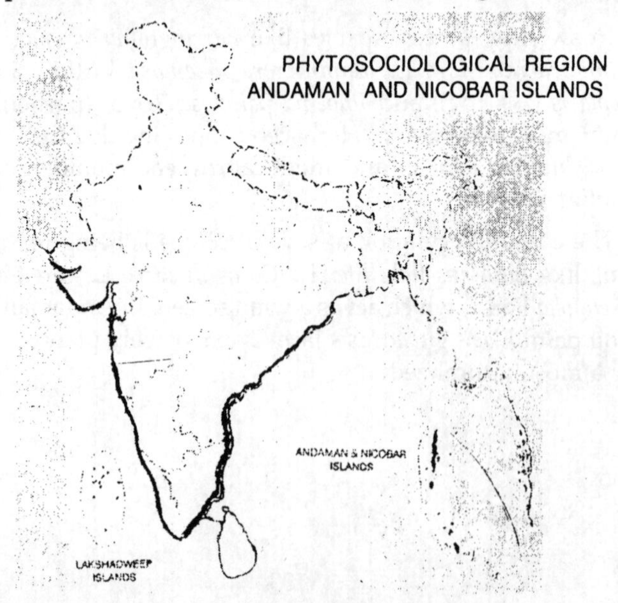

the Great Andaman (300 km long) and the Little Andaman lies to the south. The Nicobar group of Islands are separated from the mainland, from the Andaman and internally from each other by 800 m deep channel. Car Nicobar Island lies about 135 Km from the little Andaman. The Great Nicobar lies about 150 Km from Sumatra. These Islands constitute 3 groups, namely northern or the Car Nicobar, the Central and the southern or the Great Nicobar. Both Andaman and Nicobar Islands are much dissected formed of sandstones, limestones and clay of tertiary age. The highest peak in Andaman is 732 m. The higher hills are generally near the coast. The Andamans have a biogeographical affinity with Burma whereas the Nicobars with the Indonesian and south-east Asian regions. Andamans have many species of Dipterocarpaceae, characteristic of N.E. India and Burma, while Nicobars have no Dipterocarpaceae members, but show a high diversity of tree ferns and palms, characteristic of Indonesian forests.

Climate is hot and humid with the day temperature varying between 30°C-35°C. The rainfall is very high. The mean annual rainfall is 3800 mm received from both south-west and north-west monsoon from May to almost February. November has the highest rainfall in Nicobar. There is no marked winter season as such.

Vegetation

The rich natural vegetation of the Andaman and Nicobar has been studied by foresters and botanists for over a century. Our present knowledge of the vegetation and flora of Andaman and Nicobar Islands is mainly based on Kurz (1876); Parkinson (1923); Thothathri (1960a, b, 1977); Sahni (1963); Rao (1986); Balakrishnan (1989) and Saldanha (1989). As the environmental conditions and consequently the forest types and species composition differ sufficiently in Andaman and Nicobar group of Islands, they are dealt separately.

Vegetation of the Andaman group of Islands

The vegetation of these Islands can be broadly classfied as Tropical evergreen type. Balakrishnan (1989) has dealt in detail the forest types of these Islands. Based on the proximity of the sea and salinity of the soil, the vegetation can be grouped as (i) Littoral and (ii) Inland types. The littoral forests of Andamans are quite remarkable vegetation facing the open sea. They are a dense patch of evergreen vegetation comprising *Manilkara littoralis* (Kurz) Deub., *Calophyllum inophyllum* L., *Terminalia catappa* L. and *Barringtonia asiatica* (L.) Kurz. The remarkable feature of these forests is their appearance immediately bordering the open sea, just beyond the high tide mark. The littoral forests include :

(a) Submerged vegetation with marine algae such as *Enteromorpha, Ulva, Acetabularia, Codium, Caulerpa, Padina, Turbinaria and Sargassum* and sea grasses (angiosperms) like *Cymodocea, Thalassia, Halophila* and *Enhalus.*

(b) Mangrove forests with *Rhizophora mucronata* L., *R. apiculata* Bl., *Bruguiera gymnorhiza* (L.) Lam., *Ceriops tagal* (Perr.) C.B. Rob., *Avicennia marina* (Forssk.) Vierh., *Lumnitzera littorea* Voigt, *Aegiceras* corniculatus (L.) Blanc., *Excoecaria agallocha* L. and *Xylocarpus granatum* Koen.

(c) Strand vegetation : Including the dune formations and the open beach with characteristic species.

(d) Tidal or swamp forests dominated by *Cerbera odollam Gaertn., Heritiera littoralis* Dryand ex W. Ait., *Barringtonia racemosa* (L.) Spreng., *Excoecaria agallocha* L., *Sonneratia caseolaris* (L.) Engl. and a few others.

The Inland vegetation of the Andaman Islands includes the (a) Evergreen and semi-evergreen forests (b) Deciduous forests and (c) Grasslands.

Evergreen forests

Evergreen forests represent the climax vegetation and are the most magnificient among the forests of these islands. The trees are very large and straight boled and show distinct storeys. The top canopy is very dense. Humus content is poor. The tallest storey comprises gigantic trees reaching 40-60 m high. These are *Dipterocarpus grandiflorus* (Blanc.) Blanc., *D. turbinatus* Gaertn. f. *Planchonella longipetiolata* (King & Prain) *Lam., Hopea odorata* Roxb., *Endospermum diadenum* (Miq.) Airy Shaw and *Planchonia andamanica* King. The second storey is also distinct and comprises smaller trees like *Baccaurea hispida* Muell.-Arg., *Knema- cinerea* (Poir.) Warb. var. *Andamanica* (Warb.) Sinclair, *Myristica andamanica* Hk.f., *Buchanania platyneura* Kurz and *Pometia pinnata* J.R. & G. Forst.

There is little sunlight reaching the forest floor and consequently the shrubs and herbs are negligible. Among the few shrubs mention can be made of *Clerodendrum viscosum* Vent., *Leea indica* (Burm. f.) Merr., *Dinochloa scandens (*Bl. ex Nees) Kuntze (a climbing bamboo), *Maesua andamanica* Kurz, *Clinogyne grandis* Benth. & Hk.f., *Morinda umbellata* L., *Breynia vitis-idaea* (Burm. f.) Fisch. and *Grewia laevigata* Vahl. The tall trees are frequently covered by woody climbers like *Calamus longisetus* Griff., *Daemonorops mannii* Becc., *D. kurzianus* Becc. & Hk. f., *Entada pursaetha* DC., *Caesalpinia cucullata* Roxb., and *Illigera trifoliata* (Griff.) Dunn. The epiphytes are also too dense in these forests. Mainly *Asplenium nidus* L., *Drymoglossum piloselloides* (L.) Presl, *Drynaria quercifloia* (L.) Sm. and *Lycopodium phlegmaria* L. are the common pteridophytes, the epiphytic orchids belong to *Dendrobium aphyllum* (Roxb.) Fisch., *Eria* sp., and *Bulbophyllum* spp. Based on the girth and height of tree species and density of forests, the evergreen forests of Andaman can also be grouped as (1) The Giant evergreen with large trees like *Dipterocarpus costatus Gaertn. f.*, D. *gracilis* Bl. and *Calophyllum soulattri* Burm.f., (2) The Andaman tropical evergreen forests with comparatively smaller trees than the above. *Dipterocarpus grandiflorus* Bl., *Artocarpus chama* Buch.-Ham., *Planchonia andamanica* King, *Myristica andamanica* Hk. f., *Baccaurea hispida* Muell.--Arg. and climber like *Gnetum ula* Brongn. and species of *Calanus* are important.

The Andaman Semi-Evergreen Forests

These forests have both evergreen and deciduous elements mixed. The main species are *Dipterocarpus coastatus* Gaertn. f., *Pterocymbium tinctorium* (Blanc.) Merr., *Terminalia bialata* Kurz, *Artocarpus lacucha* Buch.-Ham., *Pterocarpus dalbergoioides* Roxb. ex DC. and *Dillenia pentagyna* Roxb.

The above are only a few local subtypes but broadly they are evergreen in nature.

Deciduous Forests

Deciduous forests occupy undulating hills and slopes where water holding capacity of the soil is comparatively much lower. These forests are found in North Andaman, Middle Andaman, Baratang Island and in parts of South Andaman. The dominant larger trees are *Pterocarpus dalbergioides* Roxb. ex DC. growing to almost 50 metres and with huge buttresses, *Ailanthes triphysa* (Denn.) Alst., *Parishia insignis* Hk. f., *Diploknema butyracea* (Roxb.) Lam., *Albizia lebbeck* (L.) Benth., *Tetrameles nudiflora* R. Br., *Pterocymbium tinctorum* (Blanc.) Merr., *Terminalia bilata* Kurz, *Canarium euphyllum* Kurz, *Salmalia insignis* (Wall.) Schott & Engl., *Chukrasia velutina* Wt. & Arn. are the other major species. The second storey, where distinct is characterised by smaller trees of *Sageraea elliptica* Hk. f. & Th. *Sterculia villosa* Roxb., ex. DC., *Miliusa tectona* Hutch., *Semecarpus kurzii* Engl., *Garuga pinnata* Roxb., *Cryatoxylon formosum* Benth. & Hk.f., *Pterospermum aceroides* Wall., *Lannea coromandelica* (Houtt.) Merr., *Adenanthera pavonina* L., *Dillenia pentagyna* Roxb., *Diospyros marmorata* Parker, *Semecarpus kurzii* Engl., *Cinnamomum* spp., *Grewia disperma* Rottl. ex Spreng., *Vitex glabrata* R. Br. and *Chionanthus terniflora* Wall. ex Griff. The shrubby growth is richer in this forest than in evergreen forests. Common shrubs are *Actephila excelsa* (Dalz.) Muell.-Arg., *Ixora grandiflora* Zoll. & Mor., *Licuala peltata* Roxb., *Bridelia griffithii* Hk. f., *Mallotus peltatus* (Giesel.) Muell.-Arg., and *Rinorea bengalensis* (Wall.) Gangnep. Canes and bamboos are also very frequent but sparsely distributed. Among the canes *Calamus viminalis* Willd. is the dominant one. *Gigantochloa nigrociliata* (Bruce) Kurz, *Schizostachyum kurzii* (Munro) R. Majumdar are the important bamboos. Among the climber mention can be made of *Ventilago maderaspatna* Gaertn., *Sphenodesma involucrata (Presl)* Roxb., *Pothos scandens* L., *Tetracera sarmentosa* L. and *Dinochloa scandens* (Bl. ex Nees) Kuntze spp. *andamanica* (Hoogl.) Hoogl.

The herbaceous flora is represented by grasses and the fern *Pteris quadriaurita* Retz.

Grasslands

Grasslands in Andaman are only a seral community appearing in deforested fallows. The dominant species are *Imperata cylindrica* (L.) Raeus., *Saccharum spontaneum* L., *Heteropogon contortus* P. Beauv., *Chloris barbata* Sw., *Chrysopogon aciculatus* (Retz.) Trin., and *Eragrostis* spp. These grasses are often associated with ferns like *Dicranopteris linearis* (Burm.) Underw. and *Lygodium flexuosum* (L.) Sw. and legumes like *Uraria lagopodioides* (L.) Desv. and *Desmodium heterocarpon* (L.) DC. Though rare, some shrubby species like *Melastoma malabathricum* L. and *Erycibe paniculata* Roxb. are also noticed.

Even these grasslands are now being destroyed by the invasion of the introduced adventive *Eupatorium odoratum.* L.

Vegetation of Nicobar Islands

The vegetation of Nicobar Islands is predominantly of the Nicobar type with a mixture of Malaysian and Indonesian species. The rainforests have broad-leaved multi storeyed, evergreen elements comprising of tall, straight bolled and buttressed trees with a closed canopy and supporting diverse-life forms of palms, tree ferns, epiphytes and lianas.

As in the case of Andamans, the forests of Nicobar Islands are of 2 types, namely the littoral type and the Inland type. The littoral type again comprises several subtypes like (a) Sandy beach formation and (b) Mangroves (see chapter on coast biogeographic zone). The Inland vegetation includes mainly the evergreen forests and mixed littoral forests.

The mixed littoral zone lies just behind the beach forest and stretches for some distance interior. The forest is characterised by the abundance of palms, canes and climbers. Epiphytic ferns are also profuse. The dominant trees belong to *Mangifera comptosperma* Pierre, *Terminalia procera* Roxb. *T. bialata* Steud., *Syzygium samarengense* (Bl.) Merr. & Perry, *Barringtonia racemosa* (L.) Spreng., *Heritiera microphylla* Wall. ex Kurz. Shrubby plants belong to *Tabernaemontana crispa* Roxb., *Ardisia humilis* Vahl, *Atalantia spinosa* (Willd.) Tanaka and *Hedyotis paradoxa* Kurz. Among the climbers *Calamus andamanicaus* Kurz, *Dinochloa scandens* (Bl. ex Nees) Kuntze ssp. *andamanica* (Hoogl.) Hoogl., *Thunbergia laurifolia* Lindl., *Schefflera venulosa* (Wt. & Arn.) Harms, *Mucuna gigantea* (Willd.) DC. are important. Herbaceous flora is very poor but epiphytic orchids like *Aerides emercii* Reichb. f. and *Cleisostoma uraiens* (Hayata) Garay and Sweet are frequent.

Evergreen Forests

Maximum diversity can be observed in the tropical rain forests. The trees are very robust, 40--50 m high with dense canopies and buttresses. The lianas are also many. At low altitudes the littoral and evergreen forests intermingle with each other. In such places more or less deciduous species like *Pterocymbium tinctorium* (Blanc.)

Merr., *Artocarpus chama* Buch.-Ham., *Terminalia catappa* L., *T. bialata* Steud., *T. citrina* (Gaertn.) Roxb. ex Flem., *Anthocephalus chinensis* (Lam.) A. Rich. ex Walp., *Lagerstroemia* spp., etc. are common. *Gnetum gnemon* L. also occurs scattered. Along the water courses in these forests, tree ferns like *Cyathea albo-setacea* (Bedd.) Copel. and *Angiopteris evecta* (Forst.) Hoffm. grow luxuriantly.

Pure evergreen forests occur on high and low hills and valleys in the interior of the Island. Moderate temperature and excessive rainfall have resulted in very luxuriant growth of trees here. The tall evergreen trees are *Calophyllum soulattri* (King & Prain) Lam., *Aglaia hiernii* Visw. & Ramach., *Dacryodes rugosa* H.J. Lam., *Actephila excelsa* (Dalz.) Muell.-Arg., *Fagraea auriculata* Jack., *Knema andamanica* (Warb.) de Wilde, *Litsea glutinosa* (Lour.) Robins., *Elaeocarpus serratus* L., *Nephelium lappaceum* L., *Pometia pinnata* J.R. & G. Forst., *Nothaphoebe panduriformis* (Hk. f.) Gamble and *Ficus* spp. Almost all these trees are associated with climbers, epiphytes and parasites. Among the climbers *Freycinetia sumatrana* Hemsl., Dinochloa *Scandens* (Bl. ex Nees) Kuntze, *Phanera nicobarica* Balakr. & Thoth., *Tinomiscium petiolare* Miers., *Artistolochia tagala* Cham., *Indorouchera griffithiana* (Planch.) Hall. f. are important. Epiphytes are very profuse and belong to *Aeschynanthes, Hoya, Pothos* and several orchids like *Pholidota* and *Luisia*. In some places tree ferns like *Cyathea albo-setacea* (Bedd.) Copel., Palms like *Rhopaloblaste augusta* (Kurz) Moore and *Pinanga manii* Becc. are also found in these forests.

In several places, the dense evergreen forests have been cleared for coconut plantations. *Pterocarpus dalbergioides* Roxb. ex DC., *Tectona grandis* L.f. and different species of rattans and bamboos have also been raised for economic gains.

Floristic Diversity

The Flora of Andaman group of Islands is quit rich and diverse and show striking similarities with the Malaysian flora on one hand and with Peninsular India and Sri Lanka on the other hand. Of the estimated 1500-1800 species in Andaman, 150 species are supposed to be endemic and among the non endemics about 40 per cent of the species show their extended distribution in south-east Asia including Burma, Malaya, Thailand and Sumatra but are absent in the Mainland of India (Table 13.1).

Nicobar islands are estimated to contain about 700 species of which roughly 10 per cent are endemic (Table 13.2). A number of species are reported to be extremly rare. The flora of Nicobar islands reveals a strong affinity towards the flora of Indonesia.

Although Andaman and Nicobar Islands form a chain of Islands stretching from Arakkan Yoma in Burma to Sumatra in Indonesia, the Andaman group of Islands show striking dissimilarities with the flora of Nicobar group of Islands. The genera *Dipterocarpus* and *Pterocarpus* which are widespread in Andaman are not found in the Nicobar Islands. The genera like *Otanthera* and *Astronia* (Melastomataceae), *Cyrtandra* (Gesneriaceae), *Stemonurus* (Icacinaceae), *Bentinckia* and *Rhopaloblaste* (Arecaceae) and the tree-fern *Cyathea* common in Nicobar Islands are totally absent in Andaman Islands.

The Andaman and Nicobar group of Islands are a major genetic emporium of several groups of plant resources of great economic potential. There is an enormous diversity in the timber species, of which the dominant ones are *Calophyllum inophyllum* L. 'poon', *Hertiera littoralis* Ait. 'sundri', *Pterocymbium tinctorium* (Blanc.) Merr. 'papita', *Pterygota alata* (Roxb.) R.B. 'letkak', *Elaeocarpus tuberculatus* Roxb., *Aglaia hiernii* Visw. & Ramach. 'lalchini', *Aphanamyxis polystachya* (Wall.) Parker *Dysoxylum thyrsoideum Hiern, Harpullia arborea* (Blanc.) Radik. *Pometia pinnata* Forst. f. 'thitkandu', *Lannea coromandelica* (Houtt.) Merr. 'nabba', *Mangifera comptosperma* L. 'jungli am' *Carallia brachiata* (Lour) Merr., *Terminalia bialata* Steud., *T. procera* Roxb., *Lagerstroemia ovalifolia* Teijsm & Binn., 'pyinma' *Planchonella longipetiolata* (King & Prain) Lam. 'lamba pathi' *Alstonia kurzii* Hk. f., *Premna pyramidata* Wall. ex Schauer *Knema andmanica* (Warb.) de Wilde, *Litsea glutinosa* (Lour. Robins., *Artocarpus chama* Buch.-Ham. 'toungplinne', *A gomeziana* Wall. 'lakuch', Dipterocarpus *grandiflorus* (Blanc. Blanc., *D. turbinatus* Roxb., *Hopea odorata* Roxb., *Planchonia andamanica* King, *Planchonella longipetiolata* (King & Prain) Lam. and *Pterocarpus dalbergioides* Roxb., ex DC.

Several wild relatives of tropical crop plants like *Nepheliun lappaceum* L., *Mangifera comptosperma* L. (jungli am), *Pipe betle* L., *Musa* sp., *Areca catechu* L., *Knema cinerea* (Poir. Warb. var. *andamanica* Warb.) Sincl., *Myristica elliptica* Wall. *M. andamanica* Hk. f. (wild nutmeg) also grow luxuriantly here.

Several multipurpose species like *Calamus andamanicus* Kurz, *Aglaia hiernii* Visw. & Ramach., *Rhizophora apiculata* Bl. *Dinochloa scandens* (Bl. ex Nees) Kuntze, *Vanilla andamanica* Rolfre (vanilla of commerce), *Nypa fruticans* Wurmb are well represented here.

The medicinal plants diversity is also too numerous and to mention a few *Pericampylus glaucus* (Lam.) Merr., *Sandoricum Koetijapa* (Burm. f.,) Merr., *Derris indica* (Lam.) Bennet, *Psychotria andamanica* Kurz, *P. platyneura* Kurz, *P. sarmentosa* Bl., *Ardisia oxyphylla* Wall., *Rauvolfia reflexa* Teij & Binn., *Strychnos axillaris* Colebr., *Thottea tomentosa* (Bl.) Ding Hou, *Aristolochia tagala* Cham., *Corymborkis veratrifolia* Bl., *Nervilia punctata* Makino, *Zingiber zerumbet* Sm., *Costus, speciosus* Sm, *Dioscorea glabra* Roxb. are important.

Table 13.1 : Some Endemic Taxa of Andaman and Nicobar Islands

	Species	*Endemic to Great Nicobar Islands*	*Endemic to Nicobar Islands*	*Endemic to Andaman and Nicobar Islands*
A.	***PTERIDOPHYTES***			
1.	*Cyathea albosetacea*		+	
2.	*Lindsaea tenera*			+
B.	***DICOTYLEDONS***			
3.	*Clematis smilacifolia ssp. andamanica*			+
4.	*Dillenia andamanica*			+
5.	*Artabotrys nicobarianus*		+	
6.	*Miliusa tectona*			+
7.	*Orophea katschallica*			+
8.	*Polylthia parkinsonii*			+
9.	*Pseuduvaria prainii*			+
10.	*Uvaria nicobarica*	+		
11.	*Cyclea pendulina*		+	
12.	*Tinomiscium nicobaricum*		+	
13.	Sterculia villosa			+
14.	*Glycosmis pentaphylla var. andamanensis*			+
15.	*G. pentaphylla var. insularis*			+

Contd...

Table 13.1 – *Contd...*

Species	Endemic to Great Nicobar Islands	Endemic to Nicobar Islands	Endemic to Andaman and Nicobar Islands
16. *G. pilosa*		+	
17. *Paramignya andamanica*			+
18. *Aglaia hiernii*			+
19. *Codiocarpus andamanicus*		+	
20. *Semecarpus kurzii*			+
21. *Phanera nicobarica*	+		
22. *Otanthera nicobarensis*		+	
23. *Memecylon andamanicum*		+	
24. *Hedyotis paradoxa*			+
25. *Ixora barbata*			+
26. *I. brunnescens*			+
27. *I. microsiphon*		+	
28. *I. rosella*			+
29. *Jainia nicobarica*	+		
30. *Mussaenda jelinekii*	+		
31. *Ophiorrhiza nicobarica*	+		
32. *Psychotria andamanica*			+
33. *P. platyneura*			+
34. *Tarenna weberaefolia*			+
35. *Embelia microcalyx*		+	
36. *Maesa andamanica*			+
37. *Jasminum multiflorum var. nicobaricum*		+	
38. *Alstonia kurzii*			+
39. *Chilocarpus sunainanus*	+		
40. *Tabernaemontana crispa*			+
41. *T. crispa var. nicobarica*		+	
42. *Cyrtandromoea nicobarica*		+	
43. *Cyrtandra burttii*	+		
44. *C. occidentalis*	+		
45. *Knema andamanica*			+
46. *Dehaasia candolleana*			+

Contd...

Table 13.1 – *Contd...*

Species		Endemic to Great Nicobar Islands	Endemic to Nicobar Islands	Endemic to Andaman and Nicobar Islands
47.	*Litsea kurzii*			+
48.	*Neolitsea nicobarica*		+	
49.	*Drypetes leioarpa*		+	
50.	*Euphorbia epiphylloides*			+
51.	*Glochidion calocarpum*			+
52.	*Macaranga nicobarica*		+	
53.	*Mallotus andamanicus*			+
54.	*Elatostema novarae*		+	
55.	*Pellionia procridifolia*		+	
C.	***MONOCOTYLEDONS***			
56.	*Aerides emercii*	+		
57.	*Anoectochilus nicobaricus*		+	
58.	*Eria bractescens var. kurzii*		+	
59.	*Phalaenopsis speciosa*			+
60.	*Pomatocalpa andamanica*		+	
61.	*Vanilla andamanica*			+
62.	*Hornstedtia fenzlii*		+	
63.	*Phrynium paniculatum*	+		
64.	*Dioscorea vexans*		+	
65.	*Bentinckia nicobarica*		+	
66.	*Calamus andamanicus*			+
67.	*Daemonoros kurzianus*			+
68.	*Pinanga manii*			+
69.	*Rhopaloblaste augusta*		+	
70.	*Pandanus leram*		+	
71.	*Aglaonema nicobaricum*		+	
72.	*Dinochloa scandens ssp. andamanica*		+	
	Total	**12**	**22**	**38**

Sources : Balakrishnan *et al.*, 1982.

Table 13.2 : Extra Indian Species found only in Andaman and Nicobar Islands

	Species	Extra Indian Distribution
A.	***PTERIDOPHYTES***	
1.	*Syngramma alismifolia*	Malaysia
2.	*Vittaria ensiformis*	Malaysia, Madagascar
3.	*Lindsaea parasitica*	Malaysia
4.	*L. terragona*	Malaysia, Fiji and Solomon Island.
5.	*Bolbitis sinuata*	Thailand, Malaysia,
6.	*Nephrolepis falcata*	Thailand, Vietnam, Malaysia
7.	*Davallia solida*	Burma, Malaysia, Australia.
8.	*Humata heterophylla*	Malaysia to Pacific Islands.
9.	*H. pectinata*	Sumatra to Solomon Island
10.	*Asplenium sublaserpiatiiafoliuma*	Malaya and S. China
11.	*A. tenerum var. retusum*	Malaya
12.	*Cyclosorus heterocarpus*	S. China, Malaysia
13.	*C. polycarpus*	Thailand, Malaya
14.	*Crepidomanes bilabiatum*	Thailand, W. Malaysia
15.	*Reediella humilis*	Thailand, Malaysia to New Zealand
16.	*Vandenboschia maxima*	Thailand, Vietnam, Taiwan, S. Japan, Malaysia to Pacific Islands.
17.	*Colysis macrophylla*	S. China, Malaysia, New Zealand.
18.	*C. membranacea*	Vietnam, Malaysia, New Guinea
19.	*Microsorium insigne*	Malaysia and Philippines
B.	***Angiosperms***	
20.	*Friesodielsia glauca*	Malaya and Philippines
21.	*Goniothalamus giganteus*	Thailand, Malaysia
22.	*Uvaria rufa*	Thailand, Indo-China, Malaysia to New Guinea
23.	*Tinomiscium petiolare*	Malaysia, Sumatra, Java
24.	*Casearia grewiaefolia* var. deglabrata	Malaysia to Pacific Islands.
25.	*C. fuliginosa*	Malaya, Philippines
26.	*Xanthophyllum vitellinum*	Sumatra, Java.
27.	*Saurauia bracteosa*	Java
28.	*Indorouchera griffithiana*	Malaya, Borneo, Java
29.	*Atlantia simplicifolia*	Malaya
30.	*Dacryodes rugosa*	Malaya, Borneo, Sumatra, Java.

Contd...

Table 13.2 – *Contd...*

	Species	Extra Indian Distribution
31.	*Aglaia argentea*	Malaya, Java
32.	*Dysoxylum arborescens*	Indo-China, Malaysia
33.	*D. densiflorum*	Sumatra, Java
34.	*D. macrocarpum*	Java
35.	*D. thyrosideum*	Malaya
36.	*Sandoricum koetjape*	Burma, Thailand, Indo-China, Malaysia
37.	*Smythea lanceata*	Malaysia, Java, Philippines
38.	*Allophylus dimorphus*	Indo-China, Philippines
39.	*Meliosma lanceolata*	Sumatra, Java, Borneo
40.	*Connarus semidecandrus*	Indo-China, Thailand, Malaysia to Pacific Islands
41.	*Combretum yunnanense*	China, Indo-China, Malaya, Java.
42.	*Ochthocharis javanica*	Thailand, Sumatra, Java, Borneo, Philippin.
43.	*Memecylon excelsum*	Sumatra, Java, Borneo, Malaya
44.	*Mastixia trichotoma var. maingayii*	Malaya, Java, Sumatra
45.	*Alangium javanicum*	Java
46.	*Gardenia tubifer*	Malaya, Sumatra, Java
47.	*Greenia jackii*	Malaya
48.	*Ixora kurziana*	Sumatra, Java
49.	*Mussaenda villosa*	Thailand, Malaya, Borneo
50.	*Timonius compressicaul*	Malaya, Sumatra, Java
51.	*Uncaria cordata* var. *ferruginea*	Malaya, Sumatra, Java, Borneo
52.	*Uncaria lanosa var. ferrea*	Malaya, Sumatra, Java, Borneo
53.	*Vernonia cymosa*	Java
54.	*Symplocos fasciculata*	Malaysia, Java, Philippines
55.	*Alstonia macrophylla*	Thailand, Malaysia
56.	*Rauvolfia sumatrana*	Malaya, Sumatra
57.	*Hoya wrayii*	Malaya
58.	*Aeschynanthe volubilis*	Java
59.	*Rhynchotechm parviflorum*	Malaya, Sumatra, Java
60.	*Mananthes sumatrana*	Java, Sumatra
61.	*Strobilanthes timorensis*	Timor Island
62.	*Teijsmanniodendron pteropodum*	Malaya, Indonesia, Philippines
63.	*Apama tomentosa*	Java

Contd...

Table 13.2 – *Contd...*

	Species	*Extra Indian Distribution*
64.	*Aristolochia ungulifolia*	Sumatra, Borneo
65.	*Piper clypeatum*	Malaya
66.	*Piper macropiper*	Malaya, Java
67.	*Knema laurina*	Thailand, Malaya, Borneo, Sumatra, Java
68.	*Myritica elliptica*	Malaya, Sumatra, Borneo
69.	*Kibara coriacea*	Malaya, Sumatra, Java, Celebes.
70.	*Nothophoebe panduriformi*	Malaya
71.	*Helecia serrata*	Malaya, Borneo, Sumatra, Java
72.	*Phaleria macrocarpa*	Malaya, Java
73.	*Blumeodendron kurzii*	Burma, Malaya, Java Philippines
74.	*Cleidon nitidum*	Sri Lanka
75.	*Mallotus penangensis*	Malaya, Borneo, Philippines, Sumatra
76.	*Neoscrotechinia nicobarica*	Malaya, Borneo, Philippines, Java
77.	*Spathistemon javensis*	Malaya, Borneo, Philippines, Java
78.	*Cypholophus moluccanus*	Sumatra to Pacific Islands China, Thailand, Burma, Malaysia to Pacific Islands.
79.	*Gironneira moluccanus*	China, Thailand, Burma, Malaysia to Pacific Islands.
80.	*Ficus magnoliaefolia*	Java
81.	*Apoendicula reflexa*	Thailand, Sumatra to New Guinea
82.	*Ceratostylis subulata*	Burma, Malaya, Java
83.	*Cleisostoma uraiensis*	Taiwan, Philippines
84.	*Cymbidium bicolor*	Burma, Thailand, Malaya, Indonesia
85.	*Dendrobium pensile*	Malaya
86.	*Hetaeria obliqua*	Malaya, Indonesia
87.	*H. oblongifolia*	Burma, Thailand, Malaysia, N. Australia
88.	*Nervilia punctata*	Thailand, Malaysia, Indonesia
89.	*Phalaenopsis speciosa var. tetraspis*	Java
90.	*Plocoglottis javanica*	Burma, Thailand, Malaysia, Indonesia
91.	*Podochilus microphyllus*	Burma, Thailand, Malaysia, Sumatra Java

Contd...

Table 13.2 – *Contd...*

	Species	*Extra Indian Distribution*
92.	*Pteroceras berkeleyii*	Malaya
93.	*Spathoglottis plicata*	Thailand, Indo-China, Taiwan, Malaysia to New Guinea
94.	*Thrixspermum hystrix*	Burma, Thailand, Malaysia
95.	*Trichoglottis cirrhifera*	Thailand, Malaya, Java
96.	*Vrydagzynea albida*	Bangladesh, Burma, Thailand, Indo-China, Malaysia
97.	*Korthalsia echinometra*	Burma, Malaysia
98.	*Aglaonema schottianum*	Burma, Java
99.	*Aglaonema simplex*	Java
100.	*Aglaonema simplex var. Malaccense*	Burma, Malaya
101.	*Homalomena griffithii var. ovata*	Malaya
102.	*H. nutans*	Malaysia
103.	*Pothos macrocephalus*	Malaysia
104.	*Carex Cryptostachys*	S. China, Formosa, Malaya, Philippines, New Guinea, Australia
105.	*Carex rafflesiana*	Malaysia, Java, Philippines, Australia
106.	*Mapania cuspidata* var. *angustifolia*	Malaysia
107.	*Centotheca longilamina*	Sri Lanka, Burma, Thailand, Malaya, Java
108.	*Coelorachis glandulosa*	Burma, Indo-China, Malaya, Java, Philippines

Wild species gathered for food by the local people include fruits, seeds, underground tubers, young leaves of palms, and tender shoots of bamboos and wild bananas. Fruits of *Pandanus leram* Fontane (Munkung, Nicobar bread fruit); tubers of *Dioscorea glabra* Roxb., *D. pentaphylla* L. (Yams); corms of *Colocasia esculenta* Schott and wild bananas (Musa spp.) form the major staple food of the tribal population inhabiting these Islands.

Chapter 14
PHYTO SOCIOLOGICAL REGION OF THE LAKSHADWEEP ISLANDS

The Lakshadweep Islands (Laccadives) are an archipelago of 27 small Islands stretching from 8° to 12°N latitude and 71° to 74°E longitude in the Arabian Sea. The Islands are *ca* 320 km. away from the Kerala coast. The Islands are of coral origin and most of the Islands are crescent-shaped with lagoons and coral reefs on the western side, which protect the Islands from the fierce S.W. monsoonic winds and floods. These Islands are just 0.5 to 5 m above sea level. Maximum temperature ranges between 35° to 38°C and the average annual rainfall is ca 1600 mm. Rivers and streams are absent. Only seepage water occurs below the land surface and is replenished by rain water.

The flora of these Islands has been studied mainly by Hume (1896), Prain (1889), Willis (1901), Krishnaswami (1955) and recently by Sivadasan and Joseph (1981), Sivadas *et al.*, (1983), Joseph and Madhusoodanan (1989). There is virtually no natural vegetation left. Most of the Islands are panted with coconut trees. A vast majority of the taxa now found growing in these Islands are naturalized exotics, perhaps brought by water currents, wind and migratory birds. Some of the Islands are characterised by only psammophytic herbs and shrubs. The shrubby *Scaevola and Argusia* are the only remnants of the original vegetation that can be noticed today. Some of the smaller Islands show a typical corolline rocks and beach littoral vegetation of *Pandanus-Casuarina-Thespesia* association. The shallow sea lagoons are however rich in sea grasses.

Out of the 27 Islands only 10 are inhabited and these Islands are planted with trees of *Alstonia scholaris* (L.) R. Br., *Azadirachta indica* Juss., *Calophyllum inophyllum* L., *Ficus religiosa* L., *Salmalia malabarica* (DC.) *Schott* & Lindl., *Terminalia catappa* L., *Thespesia populnea* (L.) Soland. ex Corr. and *Ziziphus mauritiana* Lam. The local inhabitants also grow several food plants like *Alocasia indica* Schott, *Annona muricata L. Artocarpus heterophyllus* Lam., *Carica papaya* L., *Citrullus lanatus* (Thunb.) Mansf., *Citrus medica* L., *Colocasia esculenta* (L.) Schott, *Cucumis sativus* L., *Cucurbita pepo* DC., *C. maxima* Duch. ex Lam., *C. moschata* Duch., *Mangifera indica* L., *Manihot esculenta* Crantz, *Moringa oleifera* Lam., *Psidium guajava* L., *Trichosanthes anguina* L. and *Zea mays* L.

The coconut plantations in almost all the Islands are invariably associated with *Areca catechu* L. and *Piper betle* L. which are the commercial crops of the Islands.

A growing Agricultural crop with coconut tree

Myristica andamanica HK.f.
(The wild nutmeg) of the Andamans Islands

A view of Hundru fall the compact clump forming bahit of the vegetation

A wild *Musa* in Andamans

Couroupita guianensis

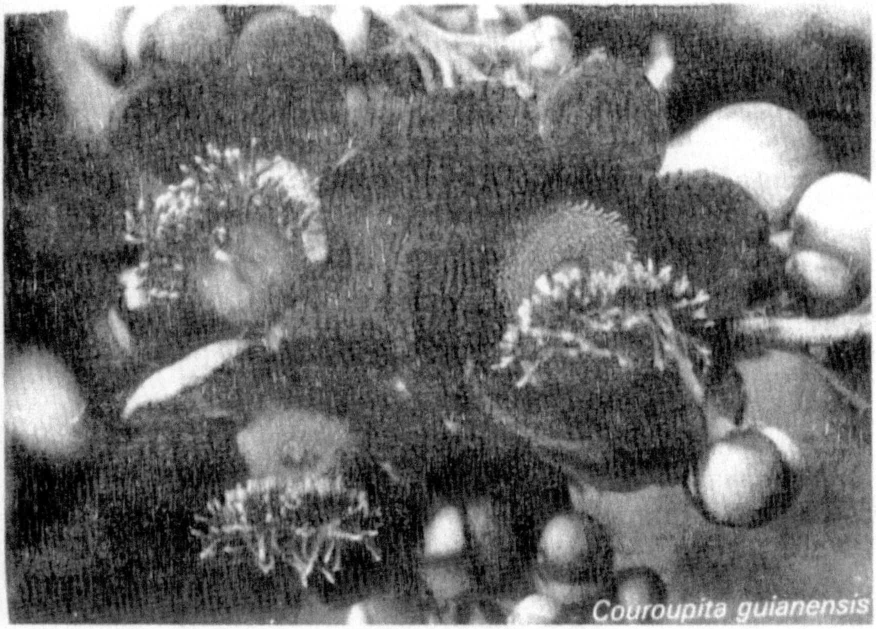

A rare Howering Plant of *Couroupita guianensis* growing in National Botanic garden, Calcutta

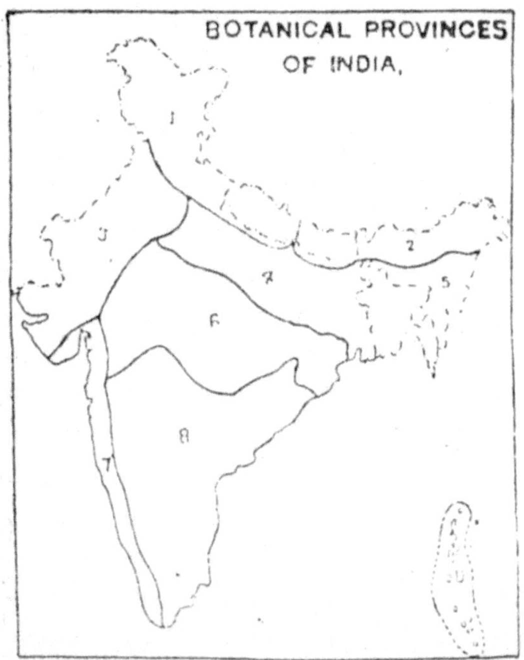

BOTANICAL PROVINCES
OF INDIA,
1 WESTERN HIMALAYAS
2 EASTERN HIMALAYAS
3 WEST INDIAN DESERT
4 GANGETIC PLAIN
5 ASSAM
6 CENTRAL INDIA
7 MALABAR
8 DECCAN
9 ANDAMANS ETC.

Map of India showing nine botanical provinces of the country

Frerea indica growing in National Botamic garden, Calcutta

Solanum grandiflorum

A rare Endangereal Plant

A *Calotrpis gigantia* growing in a desert area of Rajasthan

Butea monosperma (Lam.) Taub. Commonly known as the Flame of the forest. A common tree throughout the deciduous forests in the country

Argemone maxecoma a wild Plant & Desert area.

A view of the forest of Chotanagpur Plateau (Jharkhand)

Erimostachys superba Royle ex Benth.
(Laminaceae) a highly endengered ornamental species.

Dendrobium densiflorum Lindl. (Orchidaceae) in a rare orchid of great ornamental value.

The Desert redeves, vegetation in Rajasthan

A clump of *Dendrocalamus giganteus* Munto (Giant Bambba)

A veiw of high altitude conifer forest in Kashmir

Abies spectabilis spach.

In Himachal Pradesh (Temperate zone)

Cedrus deodaro (Roxb.) G. Don.

In Kumaon Himalaya

Delphinium cashmicrocanum Royle. On alpine herb of medicinal value.

Balanophora involucratea Hk.f.

A stemless, leafless root parasite in the tropical zone.

Anogeissus pendila Edgew - an important floristic element of the dry deciduous zone.

A view of the Tropical Evergreen forests

Rohododendron fulfgens HK.f. in Sikkim

Chapter 15

AQUATIC AND WETLAND VEGETATION

In the preceding chapters we have studied the various vegetation types prevailing in different biogeograpic zones. As the aquatic and wetland vegetation is more or less uniform in almost all zones it is preferred to discuss the same under a separate chapter.

Wetlands are transitional areas between aquatic and terrestrial ecosystem, where the water table is usually at or near the surface of the land and is covered by shallow water. They include marshes, flood plains, bogs, shallow ponds, littoral zone of the large water bodies, tidal marshes, etc. The International convention on wetland defines the wetlands as "area of marshes, fens, peatlands or water, whether natural or artificial, permanent or temporary with water, *i.e.* static or flowing, fresh, brackish or salt including areas of marine water, the depth of which at low tide does not exceed six meters". Yet another definition by IUCN (1971) reads as "submerged or water saturated land, both natural and man-made, permanent or temporary with water that is static or flowing, fresh, brackish or salt, including areas of marine water, the depth of which at low tide does not exceed six meters."

Table 15.1 : Abstracts of State/Union Territory-wise distribution of wetlands in India (including wetlands of less than 100 ha)

State		Natural		Man-Made	
		No.	Area (ha)	No.	Area (ha)
(1)		(2)	(3)	(4)	(5)
1.	Andhra Pradesh	219	1,00457	19020	4,25,892
2.	Arunachal Pradesh	2	20,200	NA	NA
3.	Assam	1394	86,355	NA	NA
4.	Bihar	62	2,24788	33	48,607
5.	Goa	3	12,360	NA	NA
6.	Gujarat	22	3,94,627	57	1,29,600
7.	Haryana	14	2,691	4	1,079
8.	Himachal Pradesh	5	702	3	19,165
9.	Jammu and Kashmir	18	7,227	NA	21,880
10.	Karnataka	10	3,320	22,758	5,39,195
11.	Kerala	32	24,329	2,121	2,10,579
12.	Madhya Pradesh	8	324	53	1,87,818
13.	Maharashtra	49	21,675	1004	2,79,025
14.	Manipur	5	26,600	NA	NA
15.	Meghalaya	2	NA	NA	NA
16.	Mizoram	3	36	1	1
17.	Nagaland	2	210	NA	NA
18.	Orissa	20	1,37,022	36	1,48,454
19.	Punjab	33	17,085	6	5,391
20.	Rajasthan	9	14,027	85	1,00,217
21.	Sikkim	42	1,101	2	3
22.	Tamil Nadu	31	58,068	20,030	2,01,132
23.	Tripura	3	575	1	4,833
24.	Uttar Pradesh	125	12,832	28	2,12,470
25.	West Bengal	54	2,91,963	9	52,564
	Total	2,167	14,58,574	65,251	25,87,965
Union Territories					
1.	Chandigarh			1	170
2.	Pondichery	3	1,533	2	1,131
	Total	**3**	**1,533**	**3**	**1,301**
	Grand Total	**2,175**	**14,60,107**	**65,254**	**25,89,266**

Source : Anonymous 1989 a.

Table 15.2 : Major Wetlands of India

	Wetland	*State*
1.	Kolleru	Andhra Pradesh
2.	Wullar	Jammu and Kashmir
3.	Chilka	Orissa
4.	Loktak	Manipur
5.	Bhoj	Madhya Pradesh
6.	Sambhar	Rajasthan
7.	Pichola	Rajasthan
8.	Ashtamudi	Kerala
9.	Harike	Punjab
10.	Ujni	Maharashtra
11.	Sukhana	Chandigarh
12.	Sasthamkotta	Kerala
13.	Renuka	Himachal Pradesh
14.	Kabar	Bihar
15.	Nalsarovar	Gujarat
16.	Kanjli	Punjab

Source : Anonymous 1989 a.

Wetlands may be natural or artificial. The former includes the fresh and salt water marshes (as in the Kutch region), rivers and streams, swamps of high rainfall region (as in Kaziranga and Manas), forested wetlands or swamp forests as Golathappar swamp forest, Mathronwala swamp forest in Doon Valley, U.P. fresh water, eustuarine, and saline lakes. The artificial wetlands include reservoirs, canals, paddy fields, fish ponds and other man-made water bodies and are areas subject to close control and careful management by man.

Wetland ecosystems are considered to be highly productive, ranging from about 300 g/m^2 in submerged macrophytes to more than 10 km/m^2 in emergent macrophytes such as *Typha angusta* Bory & Chaub., *Phragmites karka* Trin. and *Cyperus* spp. (Nair, 1989). They regulate water regime, act as natural filters, help in nutrient dynamics and act as nursery for several kinds of fish and other animals, so useful to man. As biological water purifiers *Ceratophyllum demersum* L. has been found to accumulate high concentration of chromium. *Bacopa monnierii* Penn. has been found to accumulate high quantities of copper and cadmium. *Eichhornia crassipes* (Mart.) Solms. (water hyacinth), which is considered to be a great menace to wetland ecosystem is shown to be useful as a poultry and cattle feed, as a raw material for paper industry, useful in production of biogas and as water purifiers in polluted pond by absorbing heavy metals. Therefore, conservation of wetlands has received considerable attention during the last 1 or 2 decades.

India is quite rich in wetlands which are distributed in all regions including the Cold Arid Zone of Ladakh to Warm Arid Zone of Rajasthan. Statewise distribution of wetlands (Anonymous, 1989) is shown in the Table 15.1. Some of the important wetlands are also appended in the Table 15.2.

Aquatic Vegetation

Aquatic plants can be defined as those which grow in water (at least during a major portion of the growing season) rooted or free floating, and whose growth is favoured by water logging or submergence but usually perishing during prolonged dewatering.

The aquatic and marsh vegetation of India is quite rich and diverse. Almost all the types of growth forms and life forms of aquatic plants are recognised in Indian wetlands. Approximately world's half of the aquatic flowering plants are present in the region. Of the 10 dicotyledonous and 11 monocotyledonous purely aquatic families Podostemaceae with 24 species tops the list followed by Hydrocharitaceae (14 species) and Lemnaceae (14 species). A number of aquatic plants are endemic in India, of which Podostemaceae tops the list with about 24 endemic taxa (Nagendran and Arekal, 1981).

The floristic aspect of aquatic and marsh vegetation in India is fairly well known by the numerous publications (Biswas and Calder, 1936; Subramanyam, 1962; Chavan and Sabnis, 1961; Maheshwari, 1960; Majumdar, 1965; Mirashi, 1954, 1957, 1958; Patnaik and Patnaik, 1956; Puri and Mahajan, 1958; Sarup, 1958; Unni, 1967; Satyanarayana, 1961; Sen and Chatterjee, 1959; Vyas, 1964; Bhaskar and Razi, 1973; Kachroo, 1984; Lavania *et. al.*, 1990).

The aquatic plants have been variously classified based on the growth forms in relation to the substratum, water and air (Hartog and Van der Velde, 1988). For the current discussion the broad categories as proposed by Hutchinson (1975) are considered

(i) Free Floating Forms

The free floating forms may include Pteridophytes like *Azolla pinnata* (L.) R. Br., *Salvinia natans* (L.) All., *S. auriculata* Aublet; algal members like *Chara corallina* Desv., *Nitella hyalina* (DC.) Agardh and Angiosperms like *Eichhornia crassipes* (Mart.) Solms., *Pistia stratiotes* L., *Spirodela polyrrhiza* (L.) Schleid., *Lemna aequinoctialis* Welwitsch *Wolffia microscopica* (Griff.) Kurz and *W. globosa* (Roxb.) Hartog (minute plants of Lemnaceae). Most of the free floating forms have tremendous capacity for rapid vegetative multiplication thereby chocking the entire surface of ponds. Nevertheless, they act as sinks for the nutrients added to the water body. The play a very significant role in absorption and release of heavy metals and nutrients. These carpet-forming species almost appear in pure formation. *Pistia* usually comes to profusion in November-December, the water hyacinth in December to February and *Salvinia* in August to February.

The minute plants like *Lemna, Wolffia* and *Azolla* form nice mats (Pleuston) in ponds where water is foul, usually with organic impurities. *Azolla* usually forms reddish brown cover over water surface. *Lemna* occurs in decaying ponds rich in scum usually formed by decaying diatoms or blue green algae.

(ii) Rooted Aquatics with their Foliage Floating

This group of plants is represented by hydrophyte which are rooted in the soil at the bottom of the pond with generally long petioles and scapes projecting above. The dominant species are *Nelmubo nucifera* Gaertn., *Nymphaea nouchali* Burm.f., *N. stellata* Willd., *Nymphoides indica* (L.) Kuntze and in Assam and Kashmir regions *Euryale ferox* Salisb.

These species have large, peltate leaves which just float on the surface of ponds. Commonly called 'Lotus' the *Nelumbo* and *Nymphaea* have large attractive flowers, which are considered sacred by Hindus. The other rooted aquatics particularly in shallow ponds are *Nymphoides indica* (L.) Kuntze, *Potamogeton* spp., *Aponogeton natans* Engl. & Kr. and *Trapa natans* L. var. *bispinosa* (Roxb.) Makino.

Certain species such as *Ludwigia perennis* L., *Ipomoea aquatica* Forssk., rooted in shallow ponds spread their branches floating on the surface of water.

(iii) Submerged Aquatics

Submerged aquatics include rooted hydrophytes, which are completely submerged in water. The common species are *Vallisneria spiralis* L. with long ribbon-shaped leaves and with flowers on long, filiform, coiled scapes (which are dragged down after fertilization), *Blyxa echinosperma* Hk.f., *Ottelia alismoides* Pers., *Nechamandra alternifolia* (Roxb.) Thw., *Hydrilla verticillata* Royle, *Najas graminea* Del., *Potamogeton pectinatus L.*, *P. pusillus* L., *Ceratophyllum demersum* L., *Limnophila indica* (L.) Druce. The water depth and nature of the substratuma play an important role in the production of aquatic species. Swfit flowing and deep water is not conductive for hydrophytic growth.

In the swift flowing rivers and streams on the rocks are seen firmly attached thalloid angiosperms belonging to the family Podostemaceae, chiefly represented by genera *Podostemon*, *Indotristicha*, *Dalzelia*, etc. This family of exclusively aquatic plants has maximum number (24 species) of endemic taxa in India (Nagendran *et. al.*, 1981). Although these plants are completely submerged during monsoon, when water level receds the rocks along with the plants get exposed. One can observe fine fruiting materials appearing like moss capsules at this time.

(iv) Emergent Hydrophytes

Emergent hydrophytes are those which are rooted at the bottom of the pond but with a major portion of the plant freely emerging above the surface of water. The species can withstand desiccation of pond and grow in muddy soil also. Some of the dominant ones under this catgory are *Scirpus maritimus* L., *S. articulatus* (L.) Palla, *Eleocharis palustris* (L.) R. Br., *Cyperus articulatus* L., *Juncellus serotinus* (Rottb.) Cl., *Juncus articulatus* L., *Phragmites karka* Trin., *Echinochloa crusgalli* (L.) P. Beauv., *Sacciolepis interrupta* Stapf, *Saccharum spontaneum* L. (also in terrestrial habitat), *Typha angustata* Bory & Chaub., *Limnophyton obtusifolium* Miq., *Alisma plantago-aquatica* L., *Monochoria vaginalis* (Burm. f.) Presla, *Sagittaria trifolia* L., *Butomus umbellatus* L., *Veronica angallis-aquatica* L., *Polygonum* spp., *Acorus calamus L.*, *Ranunculus sceleratus* L. and some members of Zingiberaceae. In the swampy marshes and in low moving waters, slender herbs such as *Rorippa nasturtium-aquaticum* (L.) Hayek. can be noticed almost in pure patches.

Marsh plants : Marsh plants are those which flourish on moist swampy ground, remaining saturated with water during a major portion of the year. These plants can also withstand partial inundation. Often, it is very difficult to demarcate a marsh plant with other terrertrial species which can also withstand inundation for longer or shorter duration. Marsh habitat could be fresh water marshes or saline marshes along the sea coast.

Fresh water marshes include marshy land along the banks of ponds, rivers, streams and natural swamps and even paddy fields which are inundated during some part of the year. A profusion of plant species belonging to diverse families are observed in this zone. Some of the dominant and very commonly observed plant species under this category are : *Ranunculus sceleratus* L., *Aeschynomene indica* L., *A. aspera* L., *Neptunia oleracea* Lour., *Ludwigia* spp., *Enhydra fluctuans* Lour., *Grangea maderaspatana* (L.) Poir., *Hydrolea zeylanica* Vahl., *Coldenia procumbens* L., *Limnophila spp.*, *Bacopa monnierii* Penn., *Dopatrium junceum* Ham., *Hygrophila auriculata* (Schum.) Heine, *Ammannia baccifera* L., *Bergia ammannioides* Roxb., ex Roth, *Alternanthera sessilis* (L.) R. Br. ex DC., *A. paronychioides* St.-Hill., *Polygonum orientale* L., *P. puchrum* Bl., *P. glabrum* Willd., *P. minus* Huds., *P. flaccidum* Meissn., *Monochoria vaginalis* Presl, *Xyris indica* L., *Commelina longifolia* Lam., *Juncus prismatocarpus* R. Br., *Cryptocoryne. sp.*, *Acorus calamus* L., *Caldesia parnassifolia* (L.) Parl., *Limnophyton obtusifolium* Miq., *Eriocaulon* spp., *Cyperus corymbosus* Rottb., *C. distans* L., *C. articulatus* L., *Eleocharis plantaginea* R. Br., *E. fistulosa* Link, *Scirpus articulatus* L., *Paspalidium geminatum* Stapf, *Brchiaria mutica* Stapf, *Sacciolepis interrupta* Stapf, *Panicum paludosum* Rox., *Leersia hexandra* Sw. and among pteridophytic plants *Marsilea quadrifolia* L. and *Ampelopteris prolifera* (Retz.) Copel., are important.

The salt marshes on the other hand, support altogether a different set of plant species. *Suaeda fruticosa* Forssk. ex Gmel., *S. maritima* (L.) Dunn, *S. nudiflora* Moq., *Sesuvium portulacastrum* L., *Salicornia brachiata* Roxb., *Peplidium maritimum* (L.f.) Wettst., *Arthrocnemum indicum* (Willd.) Moq., *Atriplex stocksii* Boiss. and *Acanthus illicifolius* L. are some typical ones.

Chapter 16
WEEDS AND ALIENS

A large number of alien weeds are naturalized in India and they now seem to be the permanent denizens of the native flora. The number and extent of distribution of foreign weeds in India is so great that at one time it was even thought that Indian flora is just an admixture of flora from the surrounding countries. The diversity in climate, soil types, altitude and other factors have favoured the establishment of weeds from practically all regions of the world.

Exotic weeds have been established in India ever since the time of Portuguese settlement. They introduced several economically important plants brought from Brazil, Mexico, parts of Africa and other places on their commercial route. Later on, many British officers and travellers interested in gardening also introduced many ornamental as well as medicinal plants brought from other countries to India. In the process, seeds of many obnoxious weeds also got mixed up and firmly established on the new soil. These weedy species after their introduction have spread to all parts by various factors like deforestation, shifting agriculture, faulty pasturage, sale and introduction of impure seeds, construction of roads and railway line, establishment of townships and colonies, mass shifting of labourers from one region to other for construction or for plantation work and so on. In the following paragraphs an attempt has been made to enumerate a few such adventive weeds that have naturalized in almost all botanical zones of India.

Introduced Weeds in India

Almost all the countries/biogeographic regions have contributed to the weed flora of India. The chief centres of origin of tropical weeds are Mexico, South America and Tropical Africa. A few weeds have also come from Eurasian region mainly as impurities with seeds of cultivated plants. *Stellaria media* Cyr., *Spergula arvensis* L., *Sagina apetala* L., *Anagallis arvensis* L., *Convolvulus arvensis* L., *Trigonella corniculata* L., and *Melilotus alba* Desr. are some such weeds that have invaded the cultivated fields and open fallows. While the above are a few accidental introductions, yet a few other like *Lantana camara* L. var. *aculeata* Moldenke, *Ageratum conyzoides* L., *Eupatorium adenophorum* Spreng., *Tithonia tagetiflora* Desf., *T. diversifolia* A. Gray, *Barleria cristata* L., *Euphorbia geniculata* Ort., *Jatropha gossypifolia* L., *Adhatoda zeylanica* Medic., *Pedilanthes tithymaloides* Poir., *Eichhornia crassipes* (Mart.) Solms., *Oenothera rosea* Soland., *Peperomia pellucida* H.B. & K., *Cryptostegia grandiflora* R.Br., *Opuntia elatior* Mill., *O. dillenii* Haw. etc., which were originally introduced for their ornamental value have now assumed the weedy nature and form a part of the Indian flora. The movement of inter-continental ships along with the ballst of mud infested with weeds seeds have also played a major role in the introduction and spread of many of the obnoxious weeds in India. Chiefly Asteraceae, Amaranthaceae, Papilionaceae, Malvaceae, Tiliaceae, Sterculiaceae, Rubiaceae, Acanthaceae, Euphorbiaceae, Lamiaceae, Scrophulariaceae, Cyperaceae and Poaceae are the dominant families in the weed flora. According to Maheshwari (1962) about 40 per cent of the Indian flora is introduced and now naturalized. He has classified them as (i) pluri-regional species of 'wides', (ii) weeds of cultivation, (iii) exotics and escapes from cultivation and (iv) species of limited distribution in India and adjoining regions. The weedy species that have been established all along the roadsides, railway lines, rivers, wetlands, in waste places, agriculture fields, pastures, forests are too many to list here. However, some of the dominant ones in this category are listed in the Table 16.1 (Prain, 1890; Bruhl, 1908; Kashyap, 1924; Biswas, 1934; Raizada, 1935, 36; Mathew, 1969; Mooney, 1950; Srivastava, 1955, 64; Maheshwari, 1960, 62; Maheshwari and Pal, 1976; Rao and Suryanarayana, 1979; Rao and Dam, 1979).

The neotropical weedy species, in general, have proved to be adventive in nature endangering the native flora. The Asteraceous weeds like *Eupatorium odoratum* L., *E. adenophorum* L., *Acanthospermum hispidum* DC., *Parthenium hysterophorus* L., *Mikania micrantha* H.B. & K., *Erigeron karvinskianus* DC., *Conyza bonariensis* (L.), Cronq., *Flaveria australasica* Hk., *Tithonia diversifolia* A. Gray, *T. tagetiflora* Desf., *Synedrella nodiflcra* Gaertn., *Crassocephallum crepidioides* (Benth.) S. Moore, *Xanthium strumarium* L., and *Tridax procumbens* L., in association with several other weeds like *Amaranthus spinosus* L., *Cassia tora* L., C. *occidentialis* L., *Cannabis sativa* L., *Chenaoapodium ambrosioides* L., *Nicotiana plumbaginifolia* Viv., *Alternanthera pungens* H.B. & K., *Gomphrena celosioides* Mart., *Euphorbia prostrata* Ait., E. *geniculata* Ort., *Jatropha gossypifolia* L., *Oxalis richardiana* Babu, *O. cernua* Thunb., *O. corymbosa* DC., *Scoparia dulcis* L., *Argemone mexicana* L., *Lantana camara* L., *Croton bonplandianum* Baill., *Hyptis suaveolens* Poit., *Peristrophe bicalyculata* Nees, *Calotropis* spp., and several grasses have over-run the native vegetation in many parts of India. Of these, the recently introduced *Parthenium hysterophorus* L., has become the greatest menace to Indian agriculture. This weed first reported from Poona in 1956 (Rao, 1956) has spread to all parts-agriculture fields, fallows, railway lines, roadsides, aquatic situations, forest cleared areas even forested areas. Similarly two species of *Eupatorium* , E. odoratum L., and *E. adenophorum* Spreng., have become quite adventive in comparatively open forests, fallows and roadsides on hills of Himalayan region as well as in parts of Western Ghats. At lower elevation *Mikania micrantha* H.B. & K., has become a great menace to forests of north-east India, where this weedy species climbs over large forest trees completely masking the photosynthetic surface of host plants as well as creating a nuisance in forestry operations. Another weed that has been naturalized in a remarkably short span of time is *Croton bonplandianum* Baill. This south American weed got established on Indian soil around 1900 and is now abundant in many part of India.

Among dominant weeds in the aquatic ecosystems *Eichhornia crassipes* (Mart.) Solms., *Pistia stratiotes* L., *Salvinia* spp., *Azolla pinnata* (L.) R. Br. and a few others have posed a serious threat to the native aquatic flora and fauna by completely chocking the water bodies.

Most of the wetlands in India are in grave danger of losing biological diversity. The genetic diversity in most of the aquatic species has greatly diminished over the years. Some of the wetlands which have witnessed major disruptions are mostly spread in high altitudes in the Himalayas (Salt lake, Nowgaon, Mirgund, Hokarsar, Malgaun, Sapa marshes, Apatani marshes), Arid zone (Bharatpur, Rann of Kutch, Nal Sarowar); Gangetic plains (Chaurs, Kabar jal, Harike, Sultanpur, Nawabganj), North-East India (Jaldapara, Manas, Kaziranga, Loktak lake), Deccan Peninsula (Kolleru) and in Coastal region (Rann of Kutch, Ghed swamps, Point Calimere, Pulicat lake, Bhitarkanika, Sunderbans).

Table 16.1 : Introduced Weeds in India

Species	*Family*	*Native country/Region*
Abelmoschus moschatus	Malvaceae	Paleotropical
Acalypha cilaiata	Euphorbiaceae	"
Acanthospermum hispidum	Asteraceae	Brazil
Achyranthes aspera	Amaranthaceae	Trop. America
Adathoda zeylanica	Acanthaceae	Trop. Asia
Adenostemma laevenia	Asteraceae	South America
Aeschynomene americana	Papilionacae	Trop. America
Ageratum conyzoides	Asteraceae	South America
Allamanda cathartica	Apocynaceae	Trop. America
Alternanthera ficoides	Amaranthaceae	"
A. pungens	"	"
A. sessilis	"	"
Amaranthus viridis	"	Pantropical
A. spinosus	"	"
Anagallis arvensis	Primulaceae	Europe
Antigonon leptopus	Polygonaceae	South America
Argemone mexicana	Papaveraceae	Cent. America
Asclepias curassavica	Asclepiadaceae	South America
Barleria cristata	Acanthaceae	Paleotropical
Bidens biternata	Asteraceae	America
B. pilosa	"	"
Biophytom sensitivum	Oxalidaceae	Pantropical
Blainvillea acmela	Asteraceae	Pantropical
Boerhavia diffusa	Nyctaginaceae	"

Contd...

Table 16.1 — *Contd...*

Species	*Family*	*Native country/Region*
Borreria articularis	Rubiaceae	Paleotropical
B. pusilla	"	"
Brahiaria mutica	Poaceae	Europe
Brugmontia suaveolens	Solanaceae	Mexico
Calceolaria mexicana	Scrophulariaceae	"
Canscora diffusa	Gentianaceae	Paleotropical
Cardiospermum halicacabum	Sapindaceae	Pantropical
Cassia occidentalis	Caesalpiniaceae	South America
C. pumila	"	Pantropical
C. sophera	"	South America
C. tora	"	"
Celosia argentea	Amaranthaceae	Pantropical
Ceratophyllum demersum	Ceratophyllaceae	Trop. America
Chenopodium album	Chenopodiaceae	Mexico
Chloris barbata	Poaceae	Trop. America
Chrysanthemum cinerarifolium	Asteraceae	Dalmatia
Cleome gynandra	Cleomaceae	Pantropical
C. monophylla	"	Afro-Asian
Clitoria ternatea	Papilionaceae	Paleotropical
Convolvulus arvensis	Convolvulaceae	European
Corchorus aestuans	Tiliaceae	Trop. America
C. capsularis	"	"
C. fascicularis	"	"
C. olitorius	"	Pantropical
C. tridens	"	"
C. trilocularis	"	Paleotropical
Coronopus didymus	Brassicaceae	Trop. America
Crassocephalum crepidioides	Asteraceae	"
Crotalaria medicaginea	Papilionaceae	Austro-Asian
C. verrucosa	"	Pantropical
Croton bonplandianum	Euphorbiaceae	South America
Cymbopogon martinii	Poaceae	Afro-Asian
Cynodon dactylon	"	Trop. America
Cyperus alopecuroides	Cyperaceae	Paleotropical
C. flabelliformis	"	Trop. Africa

Contd...

Table 16.1 — *Contd...*

Species	*Family*	*Native country/Region*
C. iria	"	Paleotropical
C. pumilus	"	"
C. pygmaeus	"	Pantropical
C. rotundus	"	"
C. triceps	"	Paleotropical
Dactyloctenium aegyptium	Poaceae	Pantropical
Datura stramonium	"	Paleotropical
D. metel	Solanaceae	Trop. America
Digera muricata	Amaranthaceae	Afro-Asian
Digitaria ciliaris	Poaceae	Trop. America
Drymaria cordata	Caryophyllaceae	Paleotropical
Eichhornia crassipes	Pontederiaceae	Brazil
Elephantopus scaber	Asteraceae	Pantropical
Emilia sonchifolia	"	Afro-Asian
Eragrostis cilianensis	Poaceae	"
E. tenella	"	"
Erigeron asteroides	Asteraceae	Trop. America
E. canadensis	"	South America
E. karvinskianus	"	Mexico
Eupatorium adenophorum	Asteraceae	"
E. odoratum	"	Trop. America
Euphorbia geniculata	Euphorbiaceae	Pantropical
E. prostrata	"	West Africa
E. pulcherrima	"	Mexico
Fimbristylis littoralis	Cyperaceae	Pantropical
Flaveria australasica	Asteraceae	Australia
Galinsoga quadriradiata	"	South America
G. parviflora	"	"
Glinus oppositifolius	Aizoaceae	Paleotropical
Gomphrena celosioides	Amaranthaceae	South America
Hackelochloa granularis	Poaceae	Pantropical
Heliotropium indicum	Boraginaceae	South America
H. ovalifolium	"	Pantropical
Hibiscus panduraeformis	Malvaceae	Paleotropical
Fioria vitifolius	"	"
Hypericum japonicum	Hypericaceae	"

Contd...

Table 16.1 — *Contd...*

Species	*Family*	*Native country/Region*
Hypochaeris radicata	Asteraceae	Europe
Hyptis suaveolens	Lamiaceae	South America
Ipomoeae carnea	Convolvulaceae	South America
I. eriocarpa	"	Paleotropical
I. pes-tigridis	"	"
I. reptans	"	"
Iseilema laxum	Poaceae	Trop. America
Jatropha curcas	Euphorbiaceae	"
J. glandulifera	"	Afro-Asian
J. gossypifolia	"	Trop. America
Kalanchoe pinnata	Crassulaceae	Trop. America
Laggera aurita	Asteraceae	Afro-Asian
Lantana camara L. var. aculeata	Verbenaceae	Cent. America
L. indica	"	South America
Legasceae mollis	Asteraceae	Mexico
Leucas indica	Lamiaceae	West Asia
Martynia annua	Martyniaceae	Mexico
Mecardonia dianthera	Scrophulariaceae	Trop. America
Medicago polymorpha	Papilionaceae	Eurasia/Africa
Melilotus alba	"	Eurasia
Merremia gangetica	Convolvulaceae	Asia/Africa
Mikania micrantha	Asteraceae	Trop. America
Mimosa pudica	Mimosaceae	Brazil
Murdannia dimorpha	Commelinaceae	"
Nicotiana plumbaginifolia	Solanaceae	Mexico
Nothosaerva brachiata	Amaranthaceae	Trop. Africa
Ocimum americanum	Lamiaceae	Afro-Asian
Oldenlandia corymbosa	Rubiaceae	Pantropical
Opuntia coccinellifera	Cactaceae	Mexico
O. elatior	"	South America
Ottelia alismoides	Hydrocharitaceae	Austro-Asian
Oxalis martiana	Oxalidaceae	S. America
O. richardiana	"	Mexico
Parthenium hysterophorus	Asteraceae	Trop. America
Passiflora foetida	Passifloraceae	South America
Pennisetum purpureum	Poaceae	Trop. Africa

Contd..

Table 16.1 — *Contd...*

Species	*Family*	*Native country/Region*
Peperomia pellucida	Piperaceae	Cent. America
Phyllanthus asperulatus	Euphorbiaceae	Trop. America
Physalis minima	Solanaceae	Paleotropical
P. peruviana	"	Trop. Africa
Plumbago zeylanica	Plumbaginaceae	Geront Trop.
Polycarpaea corymbosa	Caryophyllaceae	Pantropical
Polygonum barbatum	Polygonaceae	Paleotropical
P. hydropiper	"	Temperate
Portulaca oleracea	Portulacaceae	Paleotropical
Potamogeton nodosus	Potamoget-onaceae	Temperate
Pupalia lappacea	Amaranthaceae	Afro-Asian
Rhynchosia minima	Papilionaceae	Pantropical
Rivina humilis	Phytolaccaceae	South America
Ruellia tuberosa	Acanthaceae	Trop. America
Saccharum spontaneum	Poaceae	Paleotropical
Scoparia dulcis	Scrophulariaceae	South America
Sebastiania chamalea	Euphorbiaceae	Paleotropical
Sesbania bispinosa	Papilionaceae	Pantropical
Setaria glauca	Poaceae	Eurasian
S. verticillata	"	Austro-Asian
Sida alba	Malvaceae	Pantropical
S. acordifolia	"	"
S. cordata	"	Trop. America
Solanum elaeagnifolium	*Solanaceae*	Mexico
S. seaforthianum	"	Trop. America
S. sisymbrifolium	"	S. America
S. surattense	"	Paleotropical
Sonchus oleraceus	Asteraceae	"
S. wightianus	"	European
Sphaeranthus indicus	"	Africa
Sporobolus diander	Poaceae	Austro-Asian
Stachytarpheta jamaicensis	Verbenaceae	Paleotropical
Stellaria media	Caryophyllaceae	European
Synadenium grantii	Euphorbiaceae	Trop. Africa
Synedrellaa nodiflora	Asteraceae	"

Contd...

Table 16.1 — *Contd...*

Species	*Family*	*Native country/Region*
Tagetes minuta	"	"
Taraxacum officinale	"	Europe
Teramnus labialis	Papilionaceae	Pantropical
Tithonia diversifolia	Asteraceae	Mexicana
Triabulus terrestris	Zygophyllaceae	Pantropical
Tridax procumbens	Asteraceae	Mexico
Urochloa panicoides	Poaceae	Geront Trop.
Vallisneria spiralis	Hydrocharit aceae	Pantropical
Vernonia cinera	Asteraceae	"
Vicia sataiva	Papilionaceae	Eurasi/Africa
Vigna trilobata	"	Afro-Asian
Wedelia calendulacea	Asteraceae	Austro-Asian
Xanthium strumarium	"	South America
Zornia diphylla	Papilionaceae	Pantropical

Chapter 17

TAXONOMY : A VIEW

India comprises three well defined geological region : The Himalayas, the Indo-Ganga plain and the Southern Peninsula. Peninsular India is geologically old and some 110 million years ago it probably formed part of a gigantic southern continent called Gondawanaland. This was separated from a similar northern continent, called Laurasia, by the Tethys sea. Subsequently Gondwanaland broke up and parts of it now form South America, Africa, India and Australia. The change seems to have occurred after the flowering plants had come into existence so that the flora of India has some genera in common with Africa and South America. After the disintegration of Gondawanaland, the small fragement that was to form Peninsular India drifted north-east over a distance of 1,500 kilometres. The Himalayas arose later from the bed of the Tethys sea. Their uplift resulted in a deep trough between them and Peninsular India in which the alluvium of several rivers has been accumulating to give rise to the third major geological region of India, the Indo-Ganga plain, comprising nearly 700,000 square kilometres of very fertile land. The great desert of Thar in Rajasthan represents a culmination of the vast arid belt extending from the Sahara eastward through Arabia and Baluchistan.

Few other countries of comparable size possess such a rich and varied vegetation as India. A situation between 8°4′N. and 37°6′N. of Equator gives this country a great latitudinal spread which means a wide range of temperature conditions. Altitudinally, the extremes are even greater, ranging from sea level to the loftiest mountains of the world. Thus, between th plains and the mountains of India we have practically all the climatic zones from the torrid to the arctic. While in the plains the temperatures are never unfavourable for continuous plant activity, the highest peaks of the Himalayas are well above the limits of vegetation and are perpetually covered with snow. The humidity and rainfall range fom the lowest level in the desert of Rajasthan to the maximum in the hills of Assam where Cherrapunji with an annual rainfall of nearly 1,080 cm. is reputedly the rainiest spot in the world.

The rainfall in India is governed mainly by the south-west monsoon from May-June to September-October, during which we receive much of the annual rainfall, and the north-east monsoon whose influence is felt for only a short period during the rest of the year. The climate in general is also strongly influenced by the physiography of the country, particularly the following ranges of mountains and hills : (1) the Great Himalayan range with its high and low mountains, forming an almost unbroken and effective barrier on the north (2) the Western Ghats running parallel and very close to the west coast of Peninsular India together with the Nilgiris and adjoining hills of Southern India; and (3) the great complex of mountain ranges flanking the Himalayas in the north-eastern corner of the country and descending into the northern parts of Burma. Of lesser and more local influence are : (1) the Aravalli hills extending diagonally across North-western India; (2) the Vindhya and Satpura escarpment lying across Central India; and (3) the Eastern Ghats and outlying hills of Peninsular India. Stretching between the mountains under their direct of indirect influence are the Indo-Ganga plain, the semi-arid and arid plains of Western Rajasthan and adjoining areas with the Thar desert as its nucleus, the narrow coastal plains of Peninsular India, and the Deccan plateau.

2. *Botany and Plants of Ancient India*

India has been inhabited by man since almost prehistoric times. One of the earliest known civilizations flourished on Indian soil at a time when large parts of the world were still inhabited by savage men. Remains of the Indus Valley Civilization prove that rice, wheat, barley and cotton were already in cultivation in that remote period.

Many plants yielding food, fibre and wood were known to the ancient Hindus. Several others were recognized because of their real or imaginary curative properties in the treatment of the diseases of man. It is well known that even today, amongst primitive peoples, it is the tribal doctor who is better acquainted with the wild plants of the neighbourhood than other people. Early plant lore was, therefore, primarily utilitarian. Plants were classified into those which were wholesome to eat and others which were unwholesome or even injurious because of the poisons they contained. Decoctions of certain plants were used to alleviate pain, heal wounds or sores, create pleasurable excitement or act as narcotics. Early Hindu treatises like the *Ayurveda, Caraka Samhita* and *Susruta Samhita* deal with plants mainly in relation to medicine, agriculture and horticulture. The story goes that more than 2,500 years ago, Bhiksu Atreya, a well known professor at the University of Taxila, asked one of his pupils, name Jivaka, who later became the physician of King Bimbisara of Magadha, to collect, identify and describe the properties of all the plants growing within a distance of four *yojanas* from the University. Dhanvantari and Nagarjuna were other well known persons with an intimate knowledge of the characteristics of medicinal plants. *Rauvolfia,* which has now been rocketed to world-wide popularity, finds mention in ancient Hindu manuscripts as well as in the monumental work of Caraka; the plant is described under its Sanskrit name of *sarpagandha* as a useful antidote for snake bites and insect stings. Because of its curative effects in cases of insanity, it has long been known in Hindi as *pagal-ki-dawa.* The great importance of this drug has been realized only recently in western medical therapy.

The Sciences of Arboriculture, Horticulture and Silviculture were highly developed in ancient India. Methods of plant propagation by seed, cutting, layering, grafting and budding were prevalent and find mention in the Vedas, *Arthasastra* and *Brihat Samhita.* Jacolliot rightly remaked : "We should not forget that India, that immense and luminous centre in olden times, was in constant communication with all the peoples of Asia and that all the philosophers and sages of antiquity went there to study the science of life."

To a small extent, there also existed the scientific study of plants irrespective of their economic value. Branches of Botany analogous to present-day Taxonomy, Morphology, Anatomy, Physiology, Ecology, Evolution and Heredity were not wholly unknown. It appears that the ancient Hindus, like the Babylonians, had some inkling of the presence of sex in plants. For example the male plants of *Pandanus odoratissimus* L.f. were called *Ketakiviphala* or *Dhulipuspika,* the female as *Svarnaketaki,* and the male and female together as *Ketakidvayam* (or a pair of *Ketakis*). Something about the method of production of seeds in plants is also discussed in the *Harita Samhita*. Plants were classfied on the basis of their external morphology, medicinal properties and environmental associations.

In ancient India the human population was much less than what it is today. Famines were few and life was quite happy with adequate quantities of wheat, barley, rice and sugar-cane; legumes like pea, gram, *mung* and *masur*; and fruits such as mango, jack, pomegranate, date, banana, water-melon and varou limes and oranges. Among condiments, ginger, black pepper, cardamon, tamarind, onion and garlic were certainly there; and sesame was the most important source of oil, supplemented later by mustard and coconut. Of course, all the articles aof food we now know were not there in early times. Fo example, papaya, custard-apple, pineapple, guava, sapota. cashew-nut, ground-nut, maize, tomato, coffee, potato, tapioca and sweet potato were unknown at that time, having been introduced from other parts of the world in comparatively recent times. Chilli, a common flavouring material of so many of our dishes, and tobacco were also unknown.

Of the greatest importance, as an article of export, was black pepper, a native of the western coast of India. It was well known to the Greeks and was later taken to Europe by Arab traders either through the Persian Gulf, Mesopotamia and Syria, or through the Red Sea and the Gulf of Suez. At one time pepper was weighed against silver and gold, and it was the high price of pepper which acted as the chief incentive for Europeans to find a sea-route to India.

Among beverages, our ancestors had neither tea, nor coffee nor cocoa, but there were the health-giving juices of many fruits such as mango, pomegranate, coconut, citron, *jamun* and grape. Fermented drinks were also known and there was *somarasa,* the drink of the gods. It was realized early that these drinks had an exhilarating and activating effect; they were believed to increase the power of concentration and cure numerous maladies. In spite of its special virtue of giving immortality, the origin of *somarasa* remains unknown. As to the other drinks, *sidhu* was prepared from the flowers of the *mahua* tree, *kharjura* from juice of date palm, and *sura* from cereals.

As for clothing, Indians were perhaps far ahead of other countries. They were the first to weave cloth from cotton. Although cotton is now common in Egypt, this was not so 2,000 years ago. All the mummies have been found wrapped in linen, woven from the fibres of flax. In the 5th century B.C. Herodotus gave testimony that "India has wild trees tha bear fleeces as their fruits and of these the Indians make their clothes". A curious myth prevailed among the Greeks of those times that certain trees in India bore fruits which burst to produce little lambs whose soft white fleece was used to weave the finest cloth. In medieval times the muslin of Dacca was famous. It is said that this cloth could be woven so fine that a whole sheet could be folded and passed through a ring. The story goes that when the Mughal Emperor Aurangzeb once admonished his daughter for being so thinly dresed, the princes remonstrated that she was actually clad in seven folds of the finest muslin. The British finally put an end to this indigenous industry since it competed with factory-made cottom goods of British manufacture.

Among the coarser fibres were the sun-hemp, *munja*, and many others; and for stuffing, use was no doubt made of the soft silky fibre of *salmali* or *semal* (*Salmalia malabarica* Schott et Endl.) and the seed-hairs of *ak* (*Calotropis*). There is evidence however, that the Aryans knew the *Cannabis* fibre and *bhang* prepared from its leaves was often used as an intoxicant.

0Dyes of plant origin were widely used to render colour to life. The most important dye, which was also exported to other countries, was indigo–it gave a rich blue colour of great beauty and permanence. Madder obtained from the roots of a plant known as *manjistha* (*Rubia cordifolia* L.), which is quite common in the Himalayas and in South India, was used to give a bright red colour (especially popular in the *hemanta* season) and is still utilized by traders of dyeing the cloth used to bind their account books. Besides these, there were numerous flowers, wood and barks which gave a rich variety of colours, utilized by both men and women to make their dress more attractive.

Plants also featured in personal adornment and beautification of the home. *Tambula* or *pan* with all its ingredient (*kattha, supri,* cardamon etc.) was in common use to sweeten the breath. Girls were flowers of *Campaka* and jasmine in their hair and those of *siris* in their ears. They made garlands of many kinds of flowers and painted their foreheads and ckeeks with *candana* or sandal paste obtained from *Santalum album* L. Kalidasa makes frequent reference to these in his writings. The devouts used rosaries and bracelets of the seeds of *rudraksa* (*Elaecarpus ganitrus* Roxb.) or the wood of *tulasi* (*Ocimum sanctum* L).

The rich and the fastidious anointed their bodies before bathing with various fragrant pastes made of camphor, sandalwood *aguru* and the roots of *khas. Henna* or *mehndi,* a later introduction by the Muslims, was not known in those days; but women used the lac dye to colour the soles of their feet, thus reddening the flights of their steps. They also painted their lips with it and then besmeared them with a powder prepared from *lodhra* wood, which they also used as a face powder. In the *Brhat Samhita* there are references to various types of tooth-picks, hair oils, perfumes and recipes for dyeing the hair.

According to Vatsyayana, all big houses and palaces of kings had a pleasure garden - *vrksavatika* or *puspavatika* - attached to them. Season flowers like dahlia, aster, hollyhock and calendula, shrubs ike bougainvillea, and trees like eucalyptus and gold mohur were not to be found then, since all these are recent introduction. HOwever, there were othr fine trees, shrubs and climbers. Among the trees, one of the most beautiful was the red-flowered *Saraca indica* L., popularly known as the *asoka.* It is said that Sita was confined by Ravana in a grove of *asoka* trees. Another favourite tree of those days was the *kadamba* (*Anthocephalus cadamba* Miq.), whose flower appear in goden balls. It was closely connected with the life of sri Krsna and its abundance in the past near Mathura and Vrindavan is perhaps an evidence of a more humid climate in this area in those days.

Among smaller plants, *tulasi* has the pride of place and is still grown in many Hindu homes. Of climbers, *madhavilata* (*Hiptage benghalensis* Kurz) receives frequent mention in Kalidasa's plays, and among sweet-cented shrubs there were (as now) various kinds of jasmine, the musk-mallow (*Hibiscus abelmoschus* L.) and the garland-flower (*Hydychium coronarium* Koen).

Among flowers, the sacred lotus (*Nelumbo nucifera* Gaertn.) was the most important and numerous references to it occur in Sanskirt literature. According to the Puranas, Brahma emerged from the lotus which grew out of the navel of Visnu. Laksmi, the goddess of wealth and prosperity, has always been shown as standing on a lotus flower. The diurnal opening of the flower was attributed to its love for the sun, which is also responsible for the name *surya-vikaśi.* In the days of Mohenjodaro, lotus blossoms were wreathed over the head of the sun-god.

The Hindus were so fond of trees that some of them were actually deified and worshipped. Besides *asoka, padma,* and *tulasi,* the pipal and banyan were given a very high place. The tree at Budh Gaya, under which Gautama attained enlightenment, was a pipal; its branches were taken far and wide and planted to give rise to new trees. The Hindus do not readily cut the tree since it is sacred.

There are many references to the forest trees of India in Valmiki's *Ramayana.* The poet lists several plants of the Citrakuta hills, while describing the journey of Rama, Laksmana and Sita.

Chapter 18

ANGIOSPERMS*

A knowledge of the flora of India, in the modern sense, began with the zealous efforts of many European natualists and botanists who visited India in the 17th, 18th and 19th centuries, in the wake of the struggle among European nations first for trade and then for political supremacy over the country. Heinrich van Rheede, J.G. Koenig, Robert Kyd, William Roxburgh, Nathaniel Wallich, Buchanan-Hamilton, J.F. Royle, Robert Wight and J.D. Hooker were among those who laid the foundations of Botany in India. The only comprehensive flora of India so far is *A Sketch of the Flora of British India* compiled by J.D. Hooker beteween 1872 and 1897. This covers not only modern India but also Burma, Ceylon, Malaya and Pakistan, and has formed the groundwork for all the later regional floras of the country of which special mention may be made of the following : *Flora of the Presidency of Madras* by Gamble, *Flora of the Upper Gangetic Plain and of the Adjacent Siwalik and Sub-Himalayan*

* The numbers given for families, genera and species of angiosperms in the following pages are to be considered purely approximate and tentative because of the revisions that these families, genera and species have undergone since the numbers were originally published by authors like Hokker, Chatterjee, and others. This is also true of the endemic ratios. Further, these earlier authors included in their floristic treatment, regions like Burma, Ceylon and Pakistan which are adjacent to, but now politically, not parts of India.

Tracts by Duthie, *flora of Bengal* by Prain, *Flora of the Presidency of Bombay* by Cooke, *The Botany of Bihar and Orissa* by Haines, *Flora Simlensis* by Collett, and *Flora of Assam* by Kanjilal *et. al.* For a detailed bibliography on the Indian flora the reader is referred to Santapau (1958).

Hooker estimated some 174 families and 17,000 species of angiosperms and 600 species of ferns and fern allies in the flora of India. About 15,900 species of flowering plants have been described in the *Flora of British India.* As far as species are concerned, the ten dominant families are the Orchidaceae, Leguminosae, Gramineae, Rubiaceae, Euphorbiaceae, Acanthaceae, Compositae, Cyperaceae, Labiatae and Urticaceae (Including Moraceae). Excepting Labiatae and Compositae, all the rest of these ten families are more tropical than temperate. The Orchidaceae alone have a representation of more than 1,600 species in the Indian region. The greater number of Indian orchids are tropical, epiphytic and endemic. The East Himalayan region abounds in orchids, but in other parts of India they are outnumbered by the Leguminosae, Gramineae and Euphorbiaceae. The Compositae, which is the largest family in the world and predominates in the flora of many countries, takes only the seventh place in India. In fact, but for its numerous representatives in the temperate and alpine zones in the Himalayas, this family would occupy a still lower place in the list.

The ten genera each with a hundred or more species are : *Bulbophyllum, Carex, Dendrobium, Eria, Eugenia, Ficus, Habenaria, Impatiens, Pedicularis* and *Strobilanthes. Impatiens* is the largest with about 241 species distributed discontinuously in the Himalayas and the mountains of Peninsular India. Four of the genera are orchids, with *Dendrobium* (200 species) leading the list. *Eugenia, Strobilanthes** and *Pedicularis* have undergone a revision of generic limits since Hooker wrote about them.

The proportion of monocotyledons to dicotyledons is approximately 1:2:3 in genera and 1:7 in species. There are some 100 recorded species of palms and 120 species of bamboos.

---

* The genus *Strobilanthes* has since been split up into at least 25 genera even as it is related to Indian species alone. In fact, as now understood this genus has hardly a few species in India.

Endemism : It is important here to take into consideration the endemic content of the flora of India, *i.e.*, species or genera restricted in distribution to a relatively small area. The available data are, however, confined to the dicotyledons. Nevertheless, whatever findings are available on endemism are significant in that they have a direct bearing on the floristic affinity of the Indian region with the adjoining areas of the world. Two kinds of endemics are generally recognized. The palaeo-endemics or epibiotics are the surviving remnants of once successful and widespread groups and are thus on their way to extinction. The neo-endemics are new and recent forms, still in the process of extending themselves.

Islands, far removed from other areas by the ocean, show high percentages of endemic species. High mountains and very dry deserts also serve as barriers to the free spread of plants and thus bring about isolation and endemism. The high Himalayan range is effectively isolated from Northern Asia by the dry Tibetan plateau to the north and warmer alluvial plains to the south. Consequently, the temperate and alpine vegetation of the Himalayas contain several species that have been unable to migrate either north or south. Peninsular India is bounded on the north by the broad Indo-Ganga plain, and on three side by sea. Both these areas have a high endemic content. For the region as a whole (including Pakistan and Burma), 61.5 per cent of the dicotyledons are endemic. Endemic species in certain other areas are : Ceylon 30%, New Zealand 72%, Australia 80%, Hawaii Islands 82%, and California 40%. In comparison with these, 61.5% is a rather high figure for a continental area like India with land connections in three direction, east north and west. The two regions contributing most to this high endemic content are the Himalayas with 3,165 and Peninsular India with 2,045 endemic species. The number of endemic species common to both regions is 533. The Indo-Ganga plain and the desert regions of Rajasthan form an area which is extremely poor in its content of endemics. To what extent land connections between Malaysia, India and Africa have influenced the present flora of the Deccan Peninsula is difficult to indicate with any degree of precision.

The non-endemics or "wides", totalling about 38.5% of the species, are widespread and extend to other countries also. According to Chatterjee they fall into three categories : (1) those which are chiefly tropical and subtropical and of fairly wide distribution in Asia and sometimes beyond it; (2) a considerable number extending just beyond the boundaries of our area into South-western China, Thailand, Tibet and Afghanistan; (3) cultivated and introduced plants.

The dicotylendons in India are represented by 171 families. Of these, seventy-nine families (A) contain less than 20 species each; twenty-seven, (B) have 20 or more speices, of which more than 50% are non-endemics or wides; and sixty-five, (C) have 20 or more species of which more than 50% are endemics.

The families in category (A) are as follows, the figures in brackets indicating the number of species in each occurring in the Indian region : Dilleniaceae (15), Schisandraceae (5), Lardizabalaceae (5), Nymphaeaceae (11), Resedaceae (4), Bixaceaea (1), Cochlospermaceae (1), Pittosporaceae (8), Xanthophyllaceae (7), Frankeniaceaea (1), Portulacaceaae (6), Tamaricaceae (8), Elatinaceae (6), Ancistrocladaceae (5), Linaceae (8), Erythroxylaceae (6), Malpighiaceae (17), Zygophyllaceae (9), Oxalidaceae (14), Simarubaceae (15), Ochnaceae (9), Burseraceae (13), Dichapetalaceae (3), Olacaceae (18), Opiliaceae (4), Staphyleaceae (4), Hippocastanaceae (2), Sabiaceae (19), Coriariaceae (1), Droseraceae (4), Hamamelidaceae (7), Haloragaceae (14), Rhizophoraceae (16), Hernandiaceae (1), Lecythidaceae (12), Crypteroniaceae (3), Sonneratiaceae (5), Passifloraceae (7), Caricaceae (1), Turneracae (1), Datiscaceae (2), Cactaceae (6), Aizoaceae (16), Alangiaceae (6), Cornaceae (12), Nyssaceae (2), Dipsacaceae (17), Stylidaceae (3), Goodeniaceae (2), Monotropaceae (3), Diapensiaceae (1), Plumbaginaceae (8), Styracaceae (9), Salvadoraceae (5), Menyanthaceae (1), Polemoniaceae (1), Hydrophyllaceae (1), Pedaliaceae (4), Plantaginaceae (13), Nyctaginaceae (8), Illecebraceae (2), Podostemaceae (16), Nepenthaceae (1), Cytinaceae (1), Aristolochiaceae (13), Chloranthaceae (3), Myristicaceae (14), Proteaceae (7), Elaeagnaceae (12), Santalaceae (15), Balanophoraceae (6), Buxaceae (6), Ulmaceae (16), Cannabinaceae (2), Platanaceae (1), Juglandaceae (4), Myricaceae (1), Casuarinaceae (1), and Ceratophyllaceae (1).

The Dilleniaceae, Pittosporaceae and Proteaceae have their. greatest development in Australia. Along with such families as the Haloragaceae, Myristicaceae and to some extent Santalaceae, they represent the Malaysian and Australian elements in the Indian flora. The occurrence in Assam of *Nepenthes,* a genus of insectivorous plants, is interesting because this area represents the northern most limit reached bay the genus. The range of its distribution indicates a certain relationship between Madagascar and Malaysia through Ceylon and the Khasi hills of Assam. The genus *Ancistrocladus* is distributed discontinuously in West Africa on the one hand and India, Burma and Malaysia on the other.

These are two examples of a close association between the Indian and the African flora.

The Rhizophoraceae and Sonneratiaceae are predominantly mangrove families. Floristically, the Indian mangroves come under the Eastern mangrove formation extending from the east coast of Africa to Australia. The genus *Blepharistemma* is endemic to India.

The family Malpighiaceae, comprising many tropical lianas, is predominantly South American. Its presence in the Indian region (7 endemic species) and in Malaysia is remarkable; a few species are also found in Africa and Madagascar.

The Podóstemaceae are remarkable plants living on rocks in rushing water. The vegetative plant body is quite unlike that of a flowering plant and recalls an alga, moss or lichen. The family is mostly tropical, and in India its chief centre of distribution is in the south with occasional species in the Khasi hills. Of the 16 species, only 5 are wides and the rest are endemic, with 9 species endemic in South India alone.

Families like the Hamamelidaceae, Oxalidaceae, Olacaceae, Cornaceae, Dipsacaceae, Styracaceae and Elaeagnacae show the North-east Asiatic influence on the Indian flora. The Styracaceae have three centres of distribution, two of them being in America. The third, which extends from Japan to Java, touches Sikkim and the Khasi hills where, along with other species, we find a monotypic genus, *Parastyrax.* The Hamamelidaceae extend from North America through Japan and China to Sikkim and the Khasi hills. The Elaeagnaceae have a much wider distribution throughout the temperate regions and have touched the North Indian region.

The following families come under category (B) : Menispermaceae (42), Violaceae (25), Polygalaceae (32), Malvaceae (111), Sterculiaceae (80), Tiliaceae (78), Elaeocarpaceae (42),

Geraniaceae (28), Rutaceae (71), Aquifoliaceae (34), Sapindaceae (54), Connaraceae (20), Caesalpiniceae (124), Mimosaceae (96), Myrtaceae (116), Lythraceae (48), Cucurbitaceae (87), Convolvulaceae (177), Solanaceae (58), Scrophulariaceae (273), Orobanchaceae (29), Bignoniaceae (31), Verbenaceae (115), Amarantaceae (48), Chenopodiaceae (40), Thymelaeaceae (22), and Moraceae (113), Most of the members of these 27 families are distributed rather widely and also include a few temperate species.

The Violaceae, Polygalaceae and Thymelaeaceae have low endemic vlaues because of their wide distribution. Species of *Viola* occur in the mountains but *Hybanthus enneaspermus* Mull. is also met with in the plains. Species of Polygalaceae have flowers which might at first sight be genus *Daphne* is represented in the Himalayas and the Khasi hills with some six species which are all endemic. Other genera of wide distribution, which occur in the Indian region, are *Thymelaea, Edgeworthia, Wikstroemia, Stellera* and *Lasiosiphon.*

The Menispermaceae, Malvaceae, Sterculiaceae, Tiliaceae, Caesalpiniaceae, Mimosaceae, Convolvulaceae and Scrophulariaceae (in large part) form a tropical group with a wide distribution, and it is to be expected that they do not have a high percentage of endemics in any particular region of India.

In the Moraceae, the tropical genus *Ficus* with a large number of species in our area (about 86) is worthy of comment. The chief centre of development of the genus may well be Malaysia and South Burma. Our two most popular species are *Ficus religiosa* L. (pipal) and *F. benghalensis* L. (banyan), both of which are held in religious veneration and are widely planted. *Ficus krishnae* C. DC. has peculiar as cidiform leaves which, according to Hindu belief, Krishna used as cups to scoop out butter. Many species of *Ficus* begin as epiphytes on temples, old buildings, or on other plants, often between the persistent leaf bases of palms like *Phoenix*. They send down aerial roots which eventually establish contact with the soil. In this process the supporting plant is gradually killed, or the temple or building wrecked. The behaviour of the aerial roots varies. They may form buttress radiating outward in all directions, or pillars and props supporting the horizontal branches. In the epiphytic and climbing species they may clasp and surround the supporting trunk. Both the pipal and the banyan are believed to live to a great age. In some old trees the main trunk may become hollow and disintegrate, leaving the crown supported like a canopy by the numerous root pillars. The famous banyan tree at the Indian Botanical Garden, Sibpur, Calcutta, is said to be only about 2000 years in age, but has 666 aerial roots supporting a vast canopy nearly 335m. in circumference and looking more like a small forest than a single tree. Many Indian roads are lined with avenues of banyan trees stretching for several kilometres and forming an archway of shade from the hot tropical sun.

The inflorescence in *Ficus* is a syconium which ripens into the familiar fig. As is well known, the mode of pollination is extraordinary, there being a special insect (*Blastophaga,* a small wasp) adapted to *Ficus* flowers. Some species of *Ficus* show cauliflory, *i.e.,* they have their inflorescence borne directly on the trunk or older branches.

Of the Myrtaceae, the most important genus found in India is *Euagenia* (including *Syzygium* and *Jambosa*) with 103 species distributed mostly in Peninsular India. Species of *Eucalyptus* found in the hill-stations of India are all introductions from Australia which, with South America, is the chief centre of development of the Myrtaceae.

The Cucurbitaceae, Solanaceae, Amarantaceae, Chenopodiaceae and Rutaceae (in part) contain many species which have found their way to India as weeds of cultivation and have subsequently become naturalized.

The family Aquifoliaceae is represented in this country by only one genus, *Ilex,* with 34 species. It is also found in North and South America, Asia, Africa and Europe. Most of the Indian species are found also in the adjoining parts of Asia. The endemicity of *Ilex* in India is 38 per cent.

The following are the 65 families falling under the Category (C) some being tropical and others temperate : Ranunculaceae (165), Magnoliaceae (36), Annonaceae (129), Berberidaceae (35), Cruciferae (178), Fumariaceae (66), Papaveraceae (45), Capparidaceae (65), Flacourtiaceae (21), Caryophyllaceae (107), Hypericaceae (26), Guttiferae (40), Ternstroemiaceae (39), Dipterocarpaceae (51), Balsaminaceae (242), Icacinaceae (25), Meliaceae (62), Celastraceae (84), Hippocrateaceae (27),

Rhamnaceae (53), Ampelidaceae (69), Leeaceae (27), Aceraceae (20), Anacardiaceae (67), Papilionaceae (867), Rosaceae (257), Saxifragaceae (114), Crassulaceae (64), Melastomaceae (127), Combretaceae (52), Onagraceae (39), Samydaceae (20), Begoniaceae (71), Umbelliferae (180), Araliaceae (56), Caprifoliaceae (55), Rubiaceae (551),Valerianaceae (20), Compositae (696), Campanulaceae (71), Vacciniaceae (68), Ericaceae (146), Primulaceae (208), Myrsinaceae (94), Sapotaceae (32), Ebenaceae (58), Symplocaceae (51), Oleaceae (97), Apocynaceae (89), Asclepiadaceae (234), Loganiaceae (40), Gentianaceae (189), Boraginaceae (145), Lentibulariaceae (30), Gesneriaceae (133), Acanthaceae (514), Labitae (421), Polygonaceae (110), Pipercaceae (104), Lauraceae (172), Loranthaceae (73), Euphorbiaceae (444), Urticaceae (109), Cupuliferae (64), and Salicaceae (44).

Ranunculaceae (Buttercup family) : Most of the Indian representatives of this family are found in the Himalayas and the temperate regions of the Nilgiri hills. The main centre of development is in the temperate regions of the northern hemisphere. The degree of endemicity of certain genera in India is follows :- *Ranunculus* 36 per cent, *Anemone* 43 per cent, *Clematis* 76 per cent, *Thalictrum* 79 per cent, *Delphinium* 71 per cent and *Aconitum* 90 per cent. The first four of these have actinomorphic flowers and the last two have zygomorphic flowers. The low percentage of endemicity in *Ranunculus* may well be due to the weedy character of many of its members, accounting for a considerable number of wides. *Actacea spicata* L. and *Cimicifuga foetida* L. are two other plants which occur in the Himalayas and also North Asia, Europe and North America. In marked contrast to these genera are those with a restricted distribution, such as *Calathodes* occurring in the Eastern Himalayas.

Magnoliaceae (Magnolia family) : This family generally regarded by botanists as very primitive, shows a discountinuous distribution in the temperate and subtropical regions of the world. However, it comprises several endemic species. All the Inidan species of *Illicium, Talauma* and *Magnolia* are endemic; and *Manglietia* and *Michelia* show an endemicity of 80 and 73 per cent respectively. Thus, while the family itself is very old, many of the species have remained localized. Most of the Indian members of this family occur in the Eastern Himalayan and Assam regions.

Annonaceae (Custard-apple family) : The Plants belonging to this family are confined to the Tropics, especially the rain forests of Brazil, Western Africa, Ceylon, South Burma and Malaysia. However, while the members of the Old World are usually of a climbing or straggling nature and occur in dense forests, those tropical America are nearly all shrubby or arboreal and grow on open grassy plains. In India the Annonaceae are confined to the tropical parts of the Deccan and to Assam, and not a single species is found in the temperate regions of the Himalayas. Sixty per cent of the Indian species are endemic. *Polyalthia, Artabotrys* and *Annona* spp. (custard-apple, *sitaphal, ramphal*) are familiar in India. The last mentioned are introductions.

Berberidaceae (Barberry family) : The genera *Berberis* and *Mahonia* are interesting from the point of view of endemism. They extend from North Asia and Northern Europe to North America and to some extent to South America. There is a very large number of endemics, for 97 per cent of the Indian species are not found elsewhere. The general habit of *Berberis* suggests xerophytic conditions; yet in India most species of the genus are found in the humid Central and Eastern Himalayas. Very few are found in the dry North-west Himalayas. The Indian species of *Berberis* and *Mahonia* show a relationship with the Chinese species of Yunnan and adjoining areas.

Cruciferae (Mustard familiy) : This family finds its chief development in the Western Himalayas and the plains of North-west India. There are a few species in the Eastern Himalayas and the plains of North India, but in the whole of South India there are only the cultivated species and a few weeds associated with them. A much greater concentration occurs in the Mediterranean region, to which the Indian area is possibly connected through Afghanistan and Iran. Some of the common plants of this family, which are found extending from the Mediterranean region to India via West Asia, are as follows : *Matthiola odoratissima* R. Br., *Nasturtium officinale* R. Br., *Cardamine impatiens* L., and species of *Sisymbrium* and *Capsella bursa-pastoris* Medic. The total endemic percentage in India is 56, which is rather high for a widespread family like Cruciferae. Some generashow particularly high endemicity in India, e.g., *Draba* 83%, *Cardamine 70%, and* Arabis 71%. Almost all of these occur in high alpine zones.

Fumariaceae : *Corydalis* is the most important genus in the Western Himalayan and west Chinese areas. 48 species out of 61 are endemic, which brings the percentage to 79. There is evidence to suggest that the main development of *Corydalis* has taken place in Central Asia and the Himalayas, from where it has migrated east and west. A somewhat localized genus *Dactylicapnos* Wall. (syn. *Dicentra*), ranges from Kumaun to the Khasi hills and Yunnan. *Hypecoum* and *Fumaria* seem to have come to India from the West. *Fumaria indica* Pugsley is found as a weed of cultivation in several parts of India.

Papaveraceae (Poppy family) : There are 26 species of the genus *Meconopsis*, and all except one are endemic—that brings the figure of endemics to 96 per cent. *Argemone mexicana* L. (Mexican poppy), a native of tropical America, has become widely naturalized in the Indian Plains. *A. ochroleuca* Sweet sub-sp. *ochroleuca* is less common but is too has established itself in some parts of India.

Capparidaceae (Caper family) : This family, which is mainly tropical and subtropical, has a relatively smaller endemic figure of 54 per cent. *Capparis* is the largest genus with 38 species. The family is characteristic of the drier western and southern parts of the country and has a similar distribution in Africa and West Asia. *Gynandropsis, Cleome, Cadaba,* and *Crataeva* are other familiar genera.

Flacortiaceae (Flacourtia family) : The Indian members are on the whole related to the Malaysian group, except perhaps the genus *Gynocardia,* which is endemic in Sikkim and Assam. Seeds of Hydnocarpus *Kurzii* Warb yield the Chaulmoogra oil useful in the treatment of leprosy.

Caryophyllaceae (Pink family) : This family has about 57 per cent endemicity in India, chiefly in the Himalayas where it is mostly found in the temperate and alpine zones. The Mediteranean region seems to be the chief centre of distribution, and some of the species common with India through Western Asia are as follows : *Dianthus caryophyllus* L., *D. fimbriatus* Biebr., *Silene conoidea* L., *S. araneosa* C. Koch., *Stellaria aquatica* Scop., *S. bulbosa* Wulf., *Cerastium trigynum* Vill., *C. dahuricum* Fisch., *C. vulgatum* L., *Arenaria serpyllifolia* L., *Drymaria cordata* Willd., and *Polycarpaea spicta* W. et

A., *Stellaria media* Cyrill, and *Spergula arvensis* L. are two cosmopolitan weeds which occur in several parts of the country.

Guttiferae (Garcinia family) : *Poeciloneuron* is endemic in South India. *Garcinia, Calophyllum, Kayea* and *Mesua* extend from tropical Africa to Malaysia. The general endemic percentage for the family in India is 50, and most of the wides are in Malaysia. This suggests that a South-East Asian influence is responsible for the Guttiferae in our area.

Ternstroemiaceae (Tea family) : This family, sometimes called Theaceae, is discontinuously distributed in tropical Asia and tropical America. Its representatives are almost wanting in Africa and absent from Australia. The endemic percentage of the Indian species is 54. *Camellia sinensis* O. Ktze. is the tea plant of commerce extensively grown in the North-eastern and the South-western parts of the country.

Dipterocarpaceae (Sal family) : This is a tropical family characteristic of the Inde-Malayan region, but a few genera also occur exclusively in South India and Ceylon. They are very useful forest trees, the most familiar of which is the sal (*Shorea rubusta* Gaertn. f.).

Balsaminaceae (Balsam family) : As already mentioned, *Impatiens* is the largest genus of flowering plants in the Indian region, with about 241 species. The greatest concentration of the species is in the humid Eastern Himalayas and Burma; a large assemblage is also found in South India and Ceylon. *Impatiens* provide a striking example in this country of discontinuous distribution of a taxon; not a single species is common to the Himalayas and South India, Although each of these areas contains a very large number of forms. The genus as such seems to be a very old one and the two groups, the Himalayan and South Indian, must have been separated from each other for a very long time and developed in mutual isolation. The number of endemic species of *Impatiens* in India is 220 out of 241, bringing the endemic percentage to 91.

Celastraceae : This family is distributed in the lower hills and the plains of Peninsular India, Assam and the Eastern Himalayas, with a high concentration in South India. The genus *Euonymus* has 27 endemic species out of 32, bringing the endemic percentage to

84. The majority of species of *Gymnosporia* are endemic in South India and the Eastern Himalayas.

Papilionaceae (Pea family) : This is the largest family of dicotyledons in India. The total number of species is at least 867 including 372 wides and an endemic percentage of 57. The plants are of vried habit. Thus, the arborescent *Dalbergia* contrasts strontly with the small herbaceous species found in the Himalayas. *Crotalaria* and *Tephrosia* have their greatest development in South India. *Millettia* is distributed mainly in Assam and North Burma where as many as 16 species are found as endemics. *Caragana* and *Astragalus* are well developed in the dry Western Himalayas. The endemic percentage of *Astragalus* in the Himalayas is 75 and most of the species are found at high altitudes.

The family belongs predominantly to drier regions and suffers a marked diminution when we come to areas of heavy rainfall. The Assam species show a relationship with those of South-East Asia; the Himalayan, with West and North Asia; and the South and West Indian with North Africa.

Rosaceae (Rose family) : This is a family chiefly of the Northern hemisphere. In India it is mainly distributed in the temperate regions of the Himalayas and other mountains. The total number of species is 257 with 179 endemics so that the endemic percentage is 70. Most of the species are found in the alpine regions of the Himalayas, there being only a few in South India and the Indo-Ganga plain. Representative genera of the North-western Himalayas are *Prunus, Rubus, Rosa, Potentilla* and *Cotoneaster,* while those of the Eastern Himalayas are *Eribotrya Photinia* and *Pygeum.*

Saxifragaceae (Hydrangea family) : *Saxifraga* is the most important genus and is chiefly found in the temperate and alpine parts of the Eastern Himalayas. Of its 58 species, 51 are endemic, giving a percentage of 88. The genus is not found in other areas outside the Himalayas. The general endemic figure for the family in India is 76 per cent.

Rubiaceae (Coffee family) : One of the largest among the dicotyledons, this family is well represented in Peninsular India, Assam and the subtropical regions of the Himalayas. The main centre of development for our country is in South India and Ceylon, and to a lesser extent in the rain forests of Assam. The total number of Indian species is 551, of which 364 are endemic, the percentage thus coming to 67.

The coffee plant, *Coffea arabica* L., a native of Africa, is grown in the hills of South India. *Cinchona,* a native of South America, has been introduced in recent times into India and is grown in North-eastern and South India.

Compositae (Sunflower family) : This family, the largest of the flowering plants, comprises about 700 species in India, but half of this number are wides and the endemic percentage is only about 52. The plants range from the tropical region to the high alpines, and in their species content South Inida and the Himalayas are approximately equivalent.

The South Indian Compositae generally recall those of Africa, and the Himalayan recall those of North Asia and China. The Indo-Ganga plain chiefly has species of very wide distribution. Several members are grown as ornamentals. The flowers of *Carthamus tinctorius* L. (Safflower) are used in dyeing. Attempts are being made to grow *Chrysanthemum cinerariaefolium* Vis., source of the contact insecticide pyrenthrum.

Vacciniaceae (Blueberry family) : The eastern Himalayas and Assam are the richest in Vacciniaceae. Out of a total of 68 species, 64 are endemic in these areas and in Burma. *Agapetes* is the most common genus.

Ericaceae (Rhododendron family) : *Rhododendron* is a conspicuous element of the Eastern Himalayan temperate and alpine zones. A few species extend to the North-western Himalayas, to the Khasi hills and even to the Nilgiris. Out of a total of 126 species from India and Burma, 64 are endemic in the Himalayas and 44 in Upper Burma. The endemic ratio is as high as 90 per cent.

Primulaceae (Primrose family) : The two chief genera of this family, *Primula* and *Androsace,* are found only in the Himalayas. About 148 species of *Primula* seem to be endemic to the Himalayas, giving the high percentage of 91. *Androsace* is more prominent in the dry north-west of the Himalayas. *Omphalogramma* and the monotypic genus *Bryocarpum* ae confined to the Eastern Himalayas and adjoining areas of China and Burma.

Asclepiadaceae (Milkweed family) : The total number of species is estimated to be 234. Of these 172 are endemic, which gives a percentage endemic value of 73. The genus *Caralluma* is best developed in Africa and Madagascar and its occurrence in India is an example of an African element in our flora. *Ceropegia* has 40 species in India; of these 26 are endemic in South India. *Hoya,* on the other hand, is chiefly found in the Himalayas. *Utleria,* an arborescent genus with only one species, is endemic in south India. *Dischidia raffiesiana* Wall. occurs in Assam and is epiphytic like other species of this genus. *Calotropis procera* L. and C. *gigantea* L. are widespread weeds of drier areas.

Gesneriaceae : The family is found chiefly in the subtropical regions of the Eastern Himalayas, Khasi hills, Burma and Malaysia. Out of 27 genera, 7 endemic. Most of the species occur at moderate elevations in the moist hills (914 to 1,524 m.). *Didissandra* and *Aeschynanthus* are the only genera found at higher altitudes. Species of the latter often occur as epiphytes.

Acanthaceae : India is one of the richest regions for this family. There are 80 genera and 514 species occurring chiefly in the tropical and subtropical regions. They are particularly abundant in the Peninsular region, where as many as 188 species are endemic. The number of endemic genera is 14, and the general endemic figure for the family is 82 per cent. In the genus *Strobilanthes* (many of the species have recently been segregated into separate genera) 146 species are endemic out of a total of 152. Several species are characterized by a gregarious flowering at definite intervals. *S. kunthianus* T. And. (now *Phlebophyllum kunthianum* Nees) flowers once every 12 years in the Nilgiris.

Labiate (Mint family) : This is a useful family because of the volatile aromatic oils that many of its members contain. *Ocimum sanctum* L. (*tulasi*) is held in religious veneration and is commonly grown in Hindu homes. *Mentha* (*pudina,* mint) is a culinary herb prized for its flavour and aroma. Oils and perfume are obtained from *Pogostemon* (Patchouly), *Lavandula* (lavender), etc.

There are 69 genera and 421 species occurring chiefly in comparatively dry areas and moderate altitudes; of these 261 are endemic. The two chief centres of concentration are the Deccan and North-western India. The South Indian development of the Labiatae is very remarkable, recalling the Balsaminaceae. But while the latter have mostly developed in the moist Malabar region the Labiatae have multiplied in the dry eastern half of South India. The following genera show high endemicity : *Plectranthus, Anisochilus, Pogostemon, Nepeta, Leucas, Elsholtzia, Salvia, Dracocephalum, Phlomis* and *Gomphostemma. Anisochilus, Pogostemon* and *Leucas* are well represented in South India whereas *Nepeta, Elsholtzia, Salvia, Dracocephalum* and *Phlomis* aare better developed in the Himalayas.

Polygonaceae (Buckwheat family) : The genus *Polygonum* contains 88 species in the Indian area; 78 are endemic. Almost all the species found in the hills show high endemism and some of them cover a great range of altitude. *Polygonum viviparum* L. shows the greatest vertical range in distribution and is found from 1.524 to 5,484 m. In contrast to this *P. perpusillum* Hook. f. and *P. hookeri* Messin. have a very restricted range in the Himalayas.

Antigonon, Muehlenbeckia and *Coccoloba* are cultivated as ornamentals in Indian gardens. However, none of them is indigenous to this country.

Loranthaceae (Mistletoe family) : This is a family of tropical and subtropical semi-parasites. In the Old World the greatest development is seen in the Malaysian region. Of the 73 species in the Indian region. 47 are endemic. The chief concentration is in Malabar and in the moist rain forests of Assam.

Euphorbiaceae (Castor family) : Plants of this family are widely distributed in the tropical and subtropical regions of the world. The major concentration is in the Deccan Peninsula. Only 5 out of the 70 genera are endemic.

The genus *Euphorbia* shows a strong representation and 41 species are endemic out of a total of 63 which seem to be rather well balanced between Peninsular India and the Himalayas Several of the Himalayan species are found at high altitudes extending even to the alpine zone. Some of the species are fleshy and resemble cacti in their vegetative habit. However, they can be readily distinguished by their stipular thorns which occur in pairs at the nodes, as also by the presence of latex. Similar fleshy euphorbias occur in Africa.

The rubber plant, *Hevea brasiliensis* Mull.-Arg. is the most useful member of this family and a native of tropical America. It is grown on a fair scale in the Malabar region. Another useful introduction is *Manihot esculenta* (tapioca, cassava), which is also a native of tropical America. The castor, *Ricinus communis* L., is a native of tropical Africa but is now more or less naturalized in this country.

Emblica officinalis Gaertn. (*amla*), is also a very useful plant its fruits being a rich source of vitamin C. *Poinsettia* and *Codiaeum* (popularly called croton) are widely grown as ornamentals.

A comparable study of the monocotyledons is not available so far*. About 38 families are represented in India. Of these, Orchidaceae (*c.* 1,600 species), Gramineae (*c.* 750 species) and Cyperaceae (*c.* 377 species) are the most important. Further, the Orchidaceae is the largest family of flowering plants in the Indian region. The Liliaceae, Araceae and Palmaceae have about a 100 species each, or more; the Commelinaceae have about 70 and the Juncaceae about 30 species in this country. Of course, a few are very small families anywhere in the world and consist of only one genus and one species.

Hydrocharitaceae (Frogs-bit family) : There are 8 genera and 12 species in India, all being submerged aquatics. *Hydrilla, Vallisneria, Lagarosiphon, Blyxa* and *Ottelia* occur in fresh waters; *Enhalus* and *Halophila* are marine plants.

Burmanniceae : *Burmannia is* represented by about 6 species which are chiefly saprophytic. They occur mainly in South India.

Orchidaceae (Orchid family) : As already indicated. The Orchidaceae are represented by about 1,600 species in the Indian region. The largest number of species occur in the Eastern Himalayan and Assam regions. *Dendrobium, Bulbophyllum, Eria* and *Habenaria* have each about a hundred or more species in the Indian region.

The family is of great biological interest for several reasons. It is one of the largest families of flowering plants with 10,000-15,000 species widely distributed but especially abundant in the Tropics, where most of them occur as epiphytes in the evergreen rain forests. Some are saprophytes like *Neottia, Gastrodia* and *Epipogum* with fungi associated with their roots as mycorrhiza. Several of the epiphytes have aerial roots with the velamen helpful in water absorption. The flowers of orchids show remarkable adaptations for pollination by insects. Some of them are so specialized for pollination by insects. Some of them are so specialized for insects

* Chatterjee (1960) estimates that about 1,000 species of monocotyledons are endemic in the Himalayas about 500 in South India.

of certain kinds that in their absence pollination fails to occur. However, using artificial methods, several orchids belonging to different genera can be readily hybridized. Thus, *Brassocattlaelia* is a multigeneric hybrid involving three genera, *Brassavola, Cattleya* and *Laelia.*

Each fertile orchid fruit produces a dust-like mass of an enormous number of exceedingly small and light seeds which easily become wind-borne and hence, among other causes, the epiaphytic habit of so many orchids.

Because of their beautiful and gorgeous flowers, the orchids are highly prized favourites in horticulture all over the world. But the numerous wild orchids of India have not so far received, in their own country, the attention they deserve. It may be hoped that interest will be stimulated with the establishment of the National Orchidarium at Shillong.

Musaceae (Banana family) : This family belongs to the Tropics of the Old World. *Musa* is a useful genus, the banana (*Musa paradisiaca* L.) being one of the most important food plants which is widely grown in the warmer parts of the country. The cultivated varieties are usually triploid and produce no seeds. The plant is a large herb with rhizomes which are used for vegetative propagation. The inflorescence springs from the rhizome underground and emerges at the top of a false aerial stem formed mainly of the sheaths of the large leaves. The blades of the leaves are readily torn from the edges and thud wind and rain soon reduce them to a ragged condition.

Zingiberaceae (Ginger family) : The plants of this family are perennial herbs, usually with sympodial fleshy tuberous rhizomes. The family is chiefly Indo-Malayan and is well represented in the Himalayas. *Hitchenia* with 4 species is endemic to India. The rhizomes of *Curcuma* (turmeric and zeodary) and *Zingiber* (ginger) are of economic value. *Elettaria cardamomum* Maton yields the cardamom of commerce and is cultivated in the mountains of South India. The fruits of *Amomam* are also used as flavouring material. Other genera of the family found in India are : *Globba, Hedychium, Kaempferia, Alpinia, Costus* and *Roscoea.*

Cannaceae (Canna family) : *Canna,* the only genus of this family, is represented by a few species in India which are doubtfully indigenous. The family is essentially tropical and subtropical American. Several varieties of hybrid origin are garden favourities.

Marantaceae (Arrowroot family) : *Phrynium* and *Clinogyne* are the only Indian genera of this predominantly American family.

Iridaceac (Iris family) : About a dozen species of Iris, which is chiefly north temperate in distribution, occur in the Himalayas. *Crocus sativa* L. is cultivated in Kashmir for saffron, which is composed of the dried stigmas of this plant and is used for flavouring and colouring food. *Iris, Gladilus, Freesia* and *Belamcanda* are cultivated as ornamentals.

Hypoxidaceae : *Hypoxis aurea* Lour. occurs in the hills and mountains of India. There are half a dozen species of *Curculigo.*

Amaryllidaceae (Amaryllis family) : About a dozen species of Crinum and half a dozen of *Pancratium* occur in India. *Crinum, Pancratium, Zephyranthes* and *Narcissus* are cultivated in gardens.

Agavaceae : Several species of *Agave,* among them A. *Sisalana Perr,* and *A. Vera-Cruz* Mill, have been introduced and naturalized in some parts of India. Ornamental varieties of this genus as well as of *Yucca* and *Dracaena* are planted in gardens.

Taccaceae : Three species of *Tacca* occur in the country.

Dioscoreaceae (Yam family) : Some 16 species of *Discorea* and one species of *Trichopus* are the only representatives of this family in India.

Stemonaceae : Two species of *Stemona* and one of *stichoneuron* occur in India. The family as a whole consists of only three genera.

Liliaceae (Lily family) : This is a large family. There are about 160 species in India, mostly occurring in the Himalayas. Many of the species are bulbous, scapigerous herbs. Species of *Smilax* are tendril climbers. *Asphodelus tenuifolius* Cav. is a winter weed in the plains. The large orange red flowers of *Gloriosa superba* L. are so beautiful that this plant is frequently grown in gardens. Onion (*Allium cepa L.*) and garlic (*Allium Sativum* L.) are widely cultivated for their bulbs. The corms of *Colchicum autumnale* L. yeild the alkaloid colchicine. Several other plants of this family are garden favourites.

Pontederiaceae : This is a small family of aquatics, of which *Monochoria* is represented in this country by two species. *Eichhornia* *crassipes* Solms is an introduced plant now naturalized in this country.

Philydraceae : *Philydrum lanuginosum* Banks occurs in the Andaman Islands eastward to Burma and Malaysia.

Xyridaceae : Half a dozen species of *Xyris* occur in India.

Commelinaceae : This family is predominantly tropical and subtropical Nearly 70 species occur in this country; some 28 of them belong to *Aneilema* and 20 to *Commelina. Cyanotis* is another large genus of the family. *Commelina benghalensis* L., one of the commonest species, has subterranean cleistogamous flowers. *Rhoeo, Tradescantia* and *Zebrina,* all from the New world, are cultivated in gardens.

Flagellariaceae : *Flagllaria indica* L. occurs throughout India. It climbs by means of its leaf tips.

Juncaceae (Rush family) : This is a family of the temperate regions. As might be expected, it is found in India in damp and cold places, mostly in the Himalayas. There are about 26 species of *Juncus* and 4 species of *Luzula. Juncus bufonius L.* and *J. prismatocarpus* R. Br. Occur also in the plains of Northern India.

Palmaceae (Palm family) : The palms are a characteristic feature of tropical vegetation and landscape. They are of great economic importance, particularly in the Tropics. There are about 100 species of this family in India. while most of them have the distinctive palm form, some are climbers of scramblers, like *Calamus* (the largest genus of the family) and *Plectocomia.* In these the stem may attain an immense length and the leaves are provided with hooks or spines by which the stem is able to take hold of the surrounding vegetation. The rattan canes, used for making hair bottoms, baskets and cables are the stripped stems of *Calamus. Nipa fruticans* Wurmb. occurs in the tidal forests of the Sundarbans in West Bengal. *Plectocomia himalayana* Griff. and *Trachycarpus martiana* Wendl. occur up to 1,520-2,188 m. in the Eastern Himalayas. The coconut (*Cocos nucifera* L.) and the betel-nut palm (*Areca catechu* L.) are cultivated in the hot damp regions of India, the former especially near the sea. The date palm (*Phoenix dactylifera* L.) has been introduced in North-western India. The palmyra palm (*Borassus flabellifer* L.) is cultivated throughout the plains of tis country. The talipot palm (*Corypha*) is the most majestic palm of India but it is rare and confined in distribution. Many other palms are grown in gardens for ornamental purposes.

The palms are of considerable geological age. A number of palm genera are endemic with narrow ranges of distribution. The endemism of the species is even more marked; it has been estimated that 95 per cent of all the species in this family are narrowly distributed. In some of the tropical islands, all the native species are endemic and even in some continental regions the percentage is over 90. In a letter to Darwin, Hooker described the Palmaceae as "a very ancient group and much dislocated, structurally and geographically".

Pandanaceae (Screw-pine family) : They are woody plants often with a palm-like habit. The unisexual nature of the flowers of *Pandanus* seems to have been sensed by the ancient Hindus. There are about 6 species of the genus in India and one of *Freycinetia.* The sweetly scented flowers and spathes are used for ornament and as a source of perfume.

Typhaceae (Cat's-tail family) : *Typha,* the only genus of this small family, is represented by three species all grass-like marsh herbs with spongy leaves.

Sparganiaceae : This, like the above, has a sigle genus, *Sparganium.* There are two species of it in North-western India and the Eastern Himalayas.

Araceae (Aroid family) : The aroids are represented by more than a 100 species in India. The flowers are unisexual or bisexual and borne typically on a cylindrical spadix which is enclosed by a green or coloured spathe. While most of the species are terrestrial herbs, a few are climbers like *Pothos,* or epiphytes with aerial roots. *Pistia* is a floating aquatic, and *Acorus* and *Cryptocoryne* are usually marsh plants. There are several species of *Arisaema* (cobra plant, jack-in-the-pulpit) in the Himalayas and mountains of South India. *Lagenandra* with five species is endemic to South India and Ceylon. The corm-like rhizomes of *Amorphophallus, Colocasia* and *Alocasia* are edible. The rhizome of *Acorus* is aromatic, and is useful in medicine and for flavouring.

Lemnaceae (Duckweed family) : These are small, floating water plants with a thalloid plant body which is often purplish beneath. The flowers are minute and unisexual and located in grooves under the edge. The plants usually propagate themselves vegetatively and so rapidly as to cover the entire surface of the water with a dense gree carpet-like mass. *Spirodela* and *Lemna* have roots; species of *Wolffia*, the smallest of flowering plants, are rootless.

Triuridaceae : Two species of *Sciaphila* are reported, one from Kerala and the other from Assam. They are slender saprophytes with scale leaves.

Alismaceae : These are marsh or water plants of various habits. *Alisma, Limnophyton, Sagittaria* and *Wisneria* are represented by a few species in India.

Butomaceae :- Like the Alismaceae, these are water or marsh plants. *Butomus* and *Butomopsis* are two well known genera. The latter should perhaps be transferred to the Alismaceae.

Scheuchzeriaceae : Two species of *Triglochin* occur in temperate and alpine Himalayas. They are scapigerous marsh herbs with a rushlike appearance.

Aponogetonaceae : *Aponogeton*, the only genus of the family, is represented by four species of which *A. monostachyon* L. and *A. crispum* Thunb. are widespread, scapigerous plants with tuberous rootstocks.

Potamogetonaceae : They aare all partly or wholly submerged aquatic plants. *Potamogeton* is represented by about 10 species occurring in fresh waters. *Ruppia* is found in brackish water and *Cymcdoceae* is marine.

Zannichelliaceae : The single genus *Zannichellia* has two species, both cosmopolitan, and occurring in fresh or brackish waters.

Najadaceae : *Najas*, the only genus of this family, is represented by four species, but his number needs revision. They are submerged aquatic plants of fresh or brackish waters.

Eriocaulaceae : There are about 34 species of *Eriocaulon* reported from India. They are marsh or aquatic scapigerous herbs. This family is especially well developed in South America.

Cyperaceae (Sedge family) : Sedges are grass-like plants, chiefly of marshy habitats. Like grasses, they are also very widely distributed throughout the world but with a narrower ecological range, ebing a counterpart as it were of the Gramineae in rather damper conditions. More than 377 species occur in India; nearly 130 of them belong to the genus *Carex*. They occur chiefly in the Himalayas, some at very high alpine altitudes. Of *Cyperus*, which is mainly tropical an the second largest genus, there are about 60 species, and of *Fimbristylis* about 50 species. From the pith in the culms of *Cyperus papyrus* L. (paper-reed), riverside plant, the Egyptians, as early as 2400 B.C. made the ancient wirting paper, papyrus. The rhizomes and root tubers of several species of *Cyperus* are edible and the stems are used in baskerty; but, on the whole, the family is of little economic value.

Gramineae (Grass family) : One of the largest families of flowering plants, the Gramineae is also the most widely distributed in all regions of the globe. In temperate regions, the grasses are a conspicuous element of the vegetation, forming grasslands, called prairies of steppes. While most grasses are herbaceous, the bamboos are woody and may attain a size of as much as 30 m. Grasses are readily recognized but are difficult to identify botanically. There are about 750 species in India. As regards usefulness to man and animals, the pride of place should certainly go to the grasses. The following quotation from Pohl (1954) aptly summarizes their importance : "Of all the world's flowering plants, the grasses are undoubtedly the most important to man. They contribute tremendously to the earth's green mantle of vegetation they are the source of the principal foods of man and is domestic animals. Without the grasses, agriculture would be virtually impossible : grain, sugar, syrup, spices, paper, perfume, pasture, oil and timber, and a thousand other items of daily use are products of various grasses. They hold the hills, plains and mountains against the destructive erosive forces of wind and water. In the end, they form the sod that covers the sleeping dead."

Chapter 19

GYMNOSPERMS

Out of a total of about 65 living genera of gymnosperms only 14 occur in India. This is because they are mainly temperate plants and in this country only the Himalayas afford the main temperate region. Here they grow luxuriantly and some of them form extensive forests. Since the group is economically important and mostly consists of trees, it has received considerable attention at the hands of silviculturists.

Of the six living orders, four—Cycadales, Coniferales, Ephedrales and Gnetales — are represented in India, the Coniferales being the dominant order. *Ginkgo biloba* L., the sole living representative of the Ginkgoales, is Chinese although a few trees are being grown in scattered places having a mild climate.

Cycadales : *Cycas* is the only genus of this order in India. There are four species - *C. beddomei* Dyer, *C. circinalis* L., *C. pectinata* Griff. and *C. rumphii* Miq. *C. beddomei* is restricted to the dry hills of the Cuddapah Distict of Andhra Pradesh. *C. circinalis* is widely distributed in the dry deciduous forests of Southern India. Sago is extracted from its stem and seeds. *C. pectinata* is found in several parts of Eastern India and *C. rumphii* in the Andaman and Nicobar Islands. *C. revoluta* Thunb., a Japanese species, is frequently cultivated as an ornamental plant in North India. it is vegetatively propagated and generally the plants are all female.

Coniferales : The order is represented by 11 genera *Abies. Cedrus, Cephalotaxus, Cupressus, Juniperus, Larix, Picea, Pinus. Podocarpus, Taxus* and *Tsuga.* Except *Podocarpus* all the other conifers are restricted to the Himalayas. There is a clear demarcation between the West and East Himalayan conifers. There are only a few of them which occur in both regions. Their distribution is mainly governed by altitude and they grow in very characteristic formations. For example at 2,500 m. above sea level in the Western Himalayas, the tree community chiefly comprises *Cedrus deodara* Loud., *Pinus wallichiana* A. B. Jack., and *Abies pindrow* Royle along with some angiosperms like the oaks. At 1,000 m. the dominant conifer is *Pinus roxburgii* Sarg.

Abies has four species. Of these *A. pindrow* Royle and *A. spectabilis* Spach grow in the Western Himalayas and *A. densa* Griff. and *A. delavayi* Franchet in the Eastern Himalayas. They form extensive forests at high altitudes, occurring above 2,300 m. They yield a useful light wood which is used for making sleepers.

Cedrus comprises a single species, *C. deodara* Loud., occuring gregariously in the Western Himalayas at 1,200-3,300 m. At places it attains gigantic dimensions, and the oldest known tree exceeds 704 years in age. As ornamentals there are few trees in the world which compare with deodar. It is the strongest of Indian coniferous timbers being very resistant to white ants and fungi. *Cupressus torulosa* D. Don is a common associate of deodar but it does not occur in abundance.

There are two speices of *Cephalotaxus* - *C. mannii* Hk.f., and *C. griffithii* Hk.f.- both being small trees of the evergreen forests of the Eastern Himalayas.

The genus *Juniperus* comprises six species - *J. communis* L., *J. coxii* A. B. Jack., *J. macropoda* Boiss., *J. recurva* Buch.-Ham., *J. Squamata Buch.-Ham.* and *J. wallichiana* Hk.f. ex Parl. which are found in the inner valley and higher ranges, above the tree limit. They are small trees or shrubs and the twigs of many of them are burnt for incense in temples.

Larix has one species, *L. griffithiana* Hort. ex Carr. growing as a tall tree in Eastern Nepal, Sikkim and Bhutan at 2,400-3,650 m. It is the only deciduous conifer of our country.

Picea smithiana Boiss. is the West Himalayan spruce while *P. spinulosa* Henry belongs to the Eastern Himalayas. The former is a frequent associate of *Abies pindrow* Royle and attains a height of nearly 60 m. Its wood is used for cheap joinery but the bulk of the supply is utilized by the railways for treated sleepers.

Pinus is the most important genus from the point of view of forestry. Three species–*P. gerardiana* Wall. ex Lamb., *P. roxburghii* Sarg. and *P. wallichiana* A. B. Jack.–occur in the Western Himalayas and two species—*P. insularis* Endl. and *P. armandi* Franchet—are found in the Eastern Himalayas. *P. roxburghii* is put to a variety of uses. Cheap joinery in North India mostly depends on its timber. The tree is also extensively tapped for turpentine which is distilled to obtain the turpentine oil and rosin. *P. wallichiana* is an associate of *Cedrus deodara* and *Abies pindrow.* Its wood is superior to that of *P. roxburgii. P. gerardiana* grows in the dry inner valleys of the Western Himalayas and is well known for its edible seeds (*Chilgoza*) which are very nutritious.

Podocarpus neriifolius D. Don is a graceful tree occurring up to 900 m. in the Eastern Himalayas and the Andamans. It is an inhabitant of the evergreen climax forests of these regions. *P. wallichianus* Presl. has a discontinuous distribution. It occurs from Nilgiris southwards and in Assam.

Taxus baccata L. grows in moist shady places above 1,800 m. all along the Himalayas, Khasi-Jaintia hills and Naga hills. The wood is very durable but is not available in large quantities.

Tsuga dumosa Eichler is a tall pyramidal tree with gracefully drooping branches distributed in Central and Eastern Himalayas chiefly at 2,440-3,050 m. Its common associates are the spruce and the fir.

Mention may also be made here of the many introduced conifers which have now become naturalized. *Cyptomeria japonica* D. Don was introduced by seeds from Japan around the middle of the 19th century. Its cultivation was started in Darjeeling which has now extensive forests of this tree. Here it grows quickly and yields a valuable light wood. It has also become naturalized in the Western Himalayas. Cupressus funebris Endl. is a Chinese species and is frequently planted in the hills in graveyards and cemetries due to its "weeping" habit. *C. cashmeriana* Royle ex Carr. comes from

Tibet and is now widely cultivated in India. It also has a fast growth rate. *Callitris cupressiformis* Vent. was introduced from Australia into the Nilgiris in 1885. It is now the finest hedge plant of this area. *Thuja orientalis* L. has a vertical arrangement of its branches and is a common ornamental shrub in the plains of North India.

Ephedrales : *Ephedra,* the sole representative of the Ephedrales, contains erect or climbing shrubs or perennial herbs. The genus has become important in recent times because of the drug ephedrine which is extracted from some species and is used against asthma and other bronchial troubles. *E. foliata* Boiss. et Kotschy, *E. gerardiana* Wall., *E. intermedia Schrenk et Mey., E. nebrodensis Tineo, E. intermedia Schrenk et Mey., E. nebrodensis Tineo, E. regeliana Florin* and *E. saxatilis* Royle ex Florin occur in India. *E. foliata* is a scrambaling shrub found in the drier parts of Rajasthan while the other species occur at high altitudes in the Himalayas. *E. gerardiana* and *E. nebrodensis* contain good quantities of ephedrine and are being commercially exploited for this purpose.

Gnetales : The order Gnetales is represented by the genus *Gnetum* with five species–*G. contractum* Mgf., *G. gnemon* L., *G. latifolium* Bl., *G. montanum* Mgf. and *G. ula* Brongn. Of these *G. ula* is the most common. This is an extensive woody climber occurring in the evergreen forests of the Western Ghats, Andhra Pradesh and Orissa. *G. contractum* is a scandent shrub occuring in Kerala and the Nilgiri hills. *G. gnemon* is an erect shrub found in the rain forests of Eastern India. *G. montanum* grows wild in Assam, Sikkim, and parts of Orissa, and *G. latifolium* is known from the Andamans.

The gymnosperm flora of India is limited and fairly well known but details of their structure and life history have not been worked in all cases.

Chapter 20

PTERIDOPHYTES

The pteridophytes include a vast assemblage of elegant plants distributed throughout the globe, although best represented in the mountains of the Tropics. Their immense variety and very ancient lineage–going back to the Palaeozoic era–have made them a very interesting group for students of evolution.

The Himalayas, particularly the wetter eastern areas, abound in the richness and variety of ferns and their allies. As one proceeds westwards, the number of species and individuals dwindle and only a few species are found in the extreme western parts of the Himalayan range such as Kulu and Kashmir. In Rajasthan and Saurashtra the fern flora is insignificant. Along the Western Ghats, due to higher precipitation, a rich variety of ferns are met with.

In the year 1863, Beddome published a well illustrated volume on *The Ferns of Southern India,* followed by volumes entitled *Handbook to the Ferns of British India* in 1865-1870 and 1876. Later (1883), he summed up all the information in his *Handbook to the Ferns of British India, Ceylon and the Malay Peninsula,* and added a supplement in 1892. Clarke (1880) reviewed the ferns of Northern India and Hope (1899-1902) published a series of paper on *The Ferns of North-Western Himalayas.* Bllater & d'Almeida (1922) wrote on the fern flora of Bombay. Stewart (1942, 1945) gave an account of the ferns of Mussoorie, Dehra Dun and Kashmir. Mehra (1939) described the ferns of Mussoorie, and Panigrahi (1960, 1961) of

Eastern India. Alston (1948) enumerated 58 species of the genus *Selaginella* from India and 8 more have been recorded recently from the eastern parts of the country. Several workers have recently studied the ferns of Assam, Naini Tal, Mount Abu, Simla, Darjeeling and the Sikkim Himalayas.

The ferns show a wide range of habit, from the delicate filmy ferns (*Hymenophyllum, Mecodium* and *Trichomanes*) to arborescent forms (*Alsophila* and *Cyathea*) with woody trunks which sometimes attain a height of 9-12 m. or more. These tree-ferns are more common at elevations between 150 and 2,130 m. and are generally found in the shade of moist evergreen forests on mossy banks along streams.

Many ferns grow as epiphytes on trees. Among them are species of *Araiostegia, Arthromeris, Asplenium, Ctenopteris, Davallia, Drymoglossum, Drynaria, Lemmaphyllum, Lepisorus, Leptochilus, Lindsaya Loxogramme, Mecodium, Microsorium, Nephrolepis, Oleandra, Phymatodes, Polypodiu, Pyrrosia* and *Vittaria.* Of the climbing forms the most notable are *Lygodium flexuosum* Sw., L., *japonicum* Sw., L. *scandens* Sw., *Microsorium normale* Ching and *Stenochlaena palustris* Bedd.

A great many species such as *Adiantum capillus-veneris* L., *Ampelopteris prolifera* Kze., *Cyclosorus* spp., *Diplazium spp.*, *Thelypteris brunnea* Ching and *T. ciliata* Ching flourish on gravelly soil by the banks of streams. Others like *Abacopteris multilineata* Ching., *Angiopteris evecta* Hoffm., *Asplenium unilaterale* Lam. var. *rivale, Diplazium esculentum* Sw., *Osmunda regalis* L. and *Woodwardia radicans* Sm. grow along water courses. *Ceratopteris thalictroides* Brongn. is met with in tanks, streams and swampy ground and has marked aquatic adaptations. *Helminthostachys zeylancia* Hk. is seen in marshy places. *Acrostichum aureum* L. is characteristic of areas affected by tidal waters and along backwaters and is very common in most mangrove vegetations along the coast of Kerala.

Several ferns such as *Actiniopteris radiata* Link., *Adiantum caudatum* L., *Aleuritopteris albo-marginata* Panigrahi, *A. anceps* Panigrahi, *A. farinosa* Fee, *Blechnntale* L., *Cheilanthes tenuifolia* Sw., *Dryopteri crenata* Ktz., *Gleichenia glauca* Hk., *Hypodermatium crenatum* Kuhn and *Schizoloma ensifolium* J. Sm. grow on exposed rocks and show marked adaptations against drought. *Athyrium falcatum* Bedd. and *Pteridum aquilinum* Kuhn are met with in dry grassy places at higher elevations in the Western Ghats, *Schizaea digitata* Sw. occurs in the plains and at moderately high elevations in Kerala.

There is a considerable altitudinal variation in the fern flora of India. *Actiniopteris radiata* Link., *Ampelopteris prolifera* Kze., *Ceratopteris thalictroides* Brongn., *Cheilanthes tenuifolia* Sw., *Diplazium esculentum* Sw., *Drymoglossum heterophyllum* C. Chr., *Drynaria quercifoia* J. Sm., *Hemionitis arifolia* Moore, *Lygodium scandens* Sw., *Microlepia speluncae* Moore, *Nephrolepis cordifolia* Pr., *N. exaltata* Schott., *Pteris longifolia* La., *Schizoloma ensifolium* J.Sm., *Stenochlaena palustris* Bedd., *Tectaria fuscipes* C.Chr. and *T. polymorpha* Copel. are met with up to a height of 600 m. *Abacopteris multilineata* Ching, *Ampelopteris prolifera* Kze., *Araiostegia pulchra* Copel., *Asplenium lunulatum* Sw., *Cyathea spinulosa* Wall., and *Dryopteris cochleata* C.Chr. occur at 600-900 m., *Adiantum capillus-veneris* L., *Araiostegia pseudocstopteris* Copel., *A. pulchra* Coel., *Ceterach dalhousiae* C.Chr., *C. officinarum* DC., *Cyathea spinulosa* Wall., *Dicranopteris linearis* Und., *Lindsaya cultara* Sw., *Olendra musifolia* Pr., *O. neriiformis* Cav., and *Pteridum aquilinum* Kuhn at 900-1,525 m., *Adiantum venustum* D. Don, *Alsophila latebrosa* Wall., *Cheilanthese mysurensis* Wall., and *Osmunda claytoniana* L., occur at 1,525-2,135 m; *Drynaria mollis* Bedd., *Dryopteris chrysocoma* C.Chr., *Lenmaphyllum sub-rostratum* Ching and *Vittaria himalayensis* Ching at 2,135-2,750 m.; and *Leucostegia hookeri* Bedd., *Cryptogramma brunoniana* Wall., *Dryopteris barbigera* Ktz., *D. fibrillosa* C.Chr., *D. panda* C.Chr., *D. serrato-dentata* Hay. *Gymnopteris vestita* Und. and *Pteris wallichiana* Ag. at 2,750-3,660 m. At very high altitudes, between 3,660 and 4,875 m. and above occur *Cystopteris fragilis* Bernh., *Dryopteris serratodentata* Hay. *Notholaene marantae* R. Br., *Polypodium subrostratam* C. Chr., *Polystichum prescottianum* Moore and *Woodsia lanosa* Hk. In and about glaciers and glaciated beds and on rocks covered by snow at high altitudes grow *Asplenium septentrionale* Hoffm., *Cystopteris fragilis* Bernh., *Osmunda claytoniana* L. and *Polystichum lachenense* Bedd.

The aquatic and floating ferns are represented by *Azolla pinnata* R.Br., *Salvinia cucullata* Roxb. and *S. natans* All. Some species of *Marsilea* grow in ponds and ditches and along the edges of puddles and on muddy flats. *Isoetes* is generally met with in lowlands and ponds and thrives well under submerged conditions.

Psilotates : *Psilotum nudum* Beauv. is only species occurring in India. It may be terrestrial, epiphytic or saprophytic. It enjoys a fairly wide distribution in Kerala, the Nilgiri hill and the Godavari District of Andhra Pradesh. It is also known from various parts of the Deccan, West bengal, Madhya Pradesh, the Sundarbans and Assam. In the Himalayas it is found in Kumaun, extending up to Punjab and rarely to Kashmir. It is also known from the Laccadive and Minicoy Islands, and in the Barren Island it has been reported from the interior of a crater.

Lycopodiales : At least 34 species of *Lycopodium* grow in different parts of India. *L. hamiltonii* Spring., *L. phlegmaria* L., *L. phyllanthum* Hk. et Arn., *L. setaceum* Hamilt. and L. *squarrosum* Forst. are epiphytic while *L. cernuum* L., *L. clavatum* L., *L. serratum* Thunb. and *L. wightianum* Wall. are the common terrestrial forms. *L. alpinum* L. and *L. lucidulum* Michx. grow at higher elevations above 3,000 m. and *L. selago* L. reaches up to 4,900 m. *L. phlegmaria* L. is a pendulous epiphyte generally growing on trees and is also found in mangrove swamps in the Sundarbans and the Andaman Islands.

Selaginellales : About 66 species of *Selaginella* are known from India. They are generally found in moist mountainous tracts. However, some species like *S. bryopteris* Bak. and *S. repanda* Spring occur at low elevations and under xerophytic conditions. Some of these xerophytic species curl up and turn brown on drying but remarkably revive on being moistened whence the common name "resurrection plants" for such species. The better known Indian species of the genus are *S. chrysocaulos* Spring, *S. pallidissima* Spring, *S. involvens* Spring, *S. subdiaphana* Spring, *S. pentagona* Spring, *S. monospora* Spring and *S. wightii* Hieron. Generally the species are of restricted distribution and only a few like *S. subdiaphana* Spring and *S. repanda* Spring are fairly widespread. Several species have been recorded in comparatively recent years. Most of these new records are from Assam and the NEFA.

Isoetales : *Isoetes* is typically a marsh plant occurring in shallow waterlogged depressions or along the margins of ponds and pools in the drying mud. Since the plants look very much like the sedges and grasses with which they often grow associated, it is not easy to locate them in the field. *I. coromandelina* L.f. is widely distributed both in South and North India and for a long time it was the only species of *Isoetes* known from this country. However, recently five new species have been described. They are *I. sahyadrii* Mahabale and *I. dixiti* Shende from Maharashtra, *I. sampathkumarani*

L. N. Rao from Mysore, and *I. indica* Pant et Srivastava and *I. panchananii* Pant et Srivastava from Madhya Pradesh.

Equisetales : Several species of *Equisetum* occur in India; *E. arvense* L., *E. debile* Roxb., *E. diffusum* Don, *E. palustre* L. and *E. ramosissimum* Desf. Of these, *E. debile* is the most common and is met with all over India. Is variety *pashan* occurs along river banks in Poona. *E. arvense* and *E. palustre* ascend up to 3,660 m. and *E. ramosissimum* up to 3,050 m. *E. ramosissimum* var. *altissimum* grows abundantly at Mussoorie and Darjeeling at an altitude of about 1,500 m.

Ophioglossales : Several species of *Ophioglossum* occur in India : *O. aitchisonii* d'Almeida, *O. costatum* R. Br., *O. gramineum* Willd., *O. japonicum* Prantl, *O. lusitanicum* L., *O. nudicaule* L., *O. pendulum* L., *O. pedunculatum* Desv., *O. reticulatum* L. and *O. vulgatum* L. The last ascends up to 2,740 m. in Eastern Himalayas. It can be found up to 2,000 m. at Mussoorie. *O. pendulum* is an epiphytic species occurring in Kerala. Most of the other species have also been collected from Kerala.

Four species of *Botrychium* occur in mountainous parts. They are *B. daucifolium* Wall., *B. lanuginosum* Wall., *B. lunaria* Sw. and *B. ternatum* Sw. *B. lunaria* are high altitude species, occuring generally above 2,700 m. and up to 3,900 m. in Sikkim. *B. lanuginosum* reaches up to 3,000 m. in Nepal and is also common in hill-stations like Kodaikanal and Manaar in South India. The other species are met with a comparatively lower elevations, between, 1,200 m. and 2,400 m.

Helminthostachys zeylancia Hk. is met with in marshy areas in the Sundarbans, Upper Assam, Bengal and Kerala, and in the Western Ghats up to an elevation of 990 m. Recently it has also been collected from Gorakhpur in Uttar Pradesh.

Marattiales : This order is represented in India by two genera, *Angiopteris* and *Marattia*. *A. avecta* Hoffm is widely distributed at 600-1,500 m. in the Himalayas and South India, *M. fraxinea* Sm. occur in some parts of the Western Ghats at 1,200-1,800 m.

Filicales : The order Filicales includes the largest group or pteridophytes comprising several families. The Osmundaceae are represented by three species of *Osmunda* of which *O. regalis* L. is the most common. *O. cinnamomea* L. has been recorded from the

NEFA. The Schizaeaceae are represented by *Aneimia tomentosa* Sw., fou species of Lygodium, and two species of *Schizaea*. There are two genera of the Gleicheniaceae : *Gleichenia clauca* Hk. forms extensive thickets at high altitudes in North-East India; and *Dicranopteris linearsi* Und. has a wider distribution being found even in the South. The Plagiogyriaceae is a monogeneric family with *Plagiogyria euphlebia* Mett. as the more common form. The filmy-ferns (Hymenophyllaceae) are represented by several species belonging to the genera *Cephalomanes, Crepidomanes, Didymoglossum, Hymenophyllum, Mecodium, Meringium, Pleuromanes, Trichomanes* and *Vandenboschia*. Nearly all are epiphytes growing in dense rain forests on moos-covered trees trunks or moist rocks. *Mecodium polyanthos* Copel. is the most widely distributed species, ranging from 900 m. to 3,600 m. The Dicksoniaceae are represented by *Cibotium barometz*. J. Sm. in Eastern India. The Cyathaeaceae is a family of tree-ferns represented by two genera, *Alsophila* and *Cyathea;* both are arborescent forms found in the tropical rain forests of North-east India and the Western Ghats.

In the Polypodiaceae the more important genera are *Arthromeris, Cheiropleuria, Colysis, Crypsinus, Drynaria, Drymoglossum, Lemmaphyllum, Lepisorus, Leptochilus, Loxogramme, Microsoarium, Phymatodes, Platycerium, Polypodium, Pseudodrynaria* and *Pyrrosia*. They are nearly always epiphytic. The Dipteridoid group is represented by the terrestrial species *Dipteris wallichii* Moore of Assam.

The family Grammitidaceae is represented by some small epiphytes, mostly restricted to Assam. *Grammitis* and *Ctenopteris* are the more common genera. In the Thelypteridaceae, *Ampelopteris prolifera* Kze is the most widespread member of the family, apart from several species of *Cyclosorus* and *Thelypteris*. *Sphaerostephanos, Mesochlaena, Stegnogramma* and *Dictyocline* are also represented.

Among the different subfamilies of the Dennstaedtiaceae, the Dennstaedtioideae is represented by *Dennstaedtia scabra* Moore and *D. appendiculata* J. Sm.-both occurring in the Himalayas. The Lindsayoideae are represented by *Lindsaya repens* Bedd. and *L. cultrata* Sw. and species of *Scizoloma* and *Sphenomeris*. The Davallioideae include several epiphytic forms and occur mainly in Eastern India. The most common genera are *Araiostegia, Davallia,*

Davalloides, Humata and Leucostegia. Of the Oleandroideae there are three species of *Oleandra-O. wallichii* Presl., *O. musifolia* Pr. and *O. neriiformis* Cav., growing in Eastern India and 5 species of *Nephrolepis. Oleandra neriiformis occurs* in the Western Ghats also. The Pteridoideae is a large subfamily with nearly 24 genera of which *Acrostichum, Actiniopteris, Microlepia, Pteridium, Pteris* and *Stenochlaena* are the most important. A dozen species of *Bolbitis* and half a dozen of *Egenolfia* represent the Lomariopsidoideae. The Elaphoglossoideae are represented by 5 species of *Elaphoglossum.*

The Parkeriaceae is a monotypic family represented by the widespread water fern *Ceratopteris thalictroides* Brongn. which is assigned by some authors to the Gymnogrammoideae.

In the Adiantaceae the Gymnogrammoideae are represented by several species of *Adiantum, Aleuritopteris, Cheilanthese, Hemionites, Onychium* and *Pityrrogramma* all of which are fairly well distributed over the country. The Vittarioideae include two genera. *Vittaria* has about 8 species distributed in Eastern India and in Kerala while the small and epiphytic *Antrophyum* is comparatively rare. Of the Blechnoideae, *Blechnum, Doodia* and *Woodwardia* are common. In the Asplenioideae the genus *Asplenium embraces* nearly 50 species spread all over India; *Ceterach officinarum* DC. is common in the Western Himalayas. *Athyrium, Cystopteris* and *Diplazium* are the commonly met members of the Athyrioideae. In the Woodsioideae three species of *Woodsia–W. elongata* Hk., *W. alpina* Gray and *W. lanosa* Hk.—occur mostly in North-western Himalayas. Several species of *Dryopteris* and *Polystichum* represent the subfamily Dryopteridoideae. The monotypic genera *Acrophorus, Diacalpe, Lithostegia* and *Peranema* are found in the hill ranges of North-eastern India. In the Tectarioideae, *Ctenitis, Hypodermatium* and *Tectaria* are the most important.

Marsileales : Some 10 species of *Marsilea* are recorded from India. *M. minuta* L. is the commonest species, widely distributed all over India. It grows in a wide range of habitat, from higher altitudes at Mussoorie to Xeric surroundings in Bikaner. *M. brachypus* A. Br. and *M. coromandelica* Burm. occur in the Madras State. *M. ballardii* Gupta and *M. aegyptiaca* Willd. are reported from Rajasthan and *M. poonensis* Kolhat. from Maharashtra. *M. quadrifolia* L. is widely distributed in North India.

Salviniales : This order is represented by *Azolla pinnata* R.Br., *Salvinia cucullata* Roxb., and *S. natans* All. They occur in ponds and swampy fields in West Bengal, Chota Nagpur, Uttar Pradesh, Assam and other parts. Both species of *Salvinia* are widely distributed in Kerala. The plants spread quickly in the backwaters, rivers and paddy fields and are a growing menace to agriculture and water transport. *S. natans* All. occurs abundantly in the Dal Lake in Srinagar.

Economic Importance : The chief value of the Pteridophytes is in their ornamental nature. A few are recognized as useful medicinal plants. The leaves of *Adiantum caudatum* L. are said to be good for cough and fever, and those of *A. capillus-veneris* L. are administered with honey for catarrhal afflictions. The rhizomes of *A. lunulatum* L. are prescribed in fever and dysentery. Plants of *Actiniopteris radiata* Link., *Tectaria polymorpha* Copel. and the rhizome of *Blechnum orientale* L. are said to have anthelmintic properties. *Hrynaria quercifolia* J. Sm., and epiphytic fern, is believed to be useful in typhoid fever hectic fever and cough. The rhizomes and leaf bases of *Dryopteris chrysocoma* C. Chr. are a good substitute for *D. filix-mas* Schott. for use as a taenifuge, and species of *Lygodium–L. flexuosum* Sw. and *L. japonicum* Sw.–as expectorants. The roots of *L. flexuosum*, boiled with mustard oil, are good for rheumatism sprains, eczema and cut wounds. Osmunda regalis L. is used as a tonic and styptic. Recently, marsiline, a sedative and anti-convulsant principle, has been isolated from the leaves and whole plants of *Marsilea minuta* L. An oil prepared from *Ophioglossum vulgatum* L. is useful for the treatment of wounds and haemorrhages. An extract from the plants of *Lycopodium clavatum* L. is used as a kidney stimulant and that from *Equisetum arvense* L. as a diuretic.

Chapter 21

BRYOPHYTES

The Bryophytes include two major groups : the Hepaticae (liverworts) and the Musci (mosses). The mosses are the more numerous and more widely distributed than the liverworts, and are especially conspicuous in colder areas where they often constitute a prominent feature of the vegetation. However, both liverworts and mosses are at their optimum in the broad-leaved, temperate forests at an altitude of 2,000-2,500 m. Places like Dalhousie, Simla, Mussoorie, Darjeeling, Shillong and Ootacamund are very rich in their bryophytic flora. As we go higher up there is a fall in the number of species.

Hepaticae : The liverworts are of two types thallose and foliose. A large number of the thallose forms occur in moist places, some occur on exposed slopes, and a few are aquatic, *Ricciocarpus natans* Corda is an interesting floating liverwort of the Dal lake in Kashmir and has also been collected from Naini Tal and Kapurthala. *Ricci fluitans* L. has been collected from the Khajiar lake (near Dalhousie), Darjeeling and from South India. The foliose or leafy forms are generally common in shady and moist places on rocks, and some grow as epiphytes. The epiphyllous forms are common in Cherrapunji and some other humid regions in Assam and in South India. Some new genera of liverworts discovered from the Western Himalayas are *Aitchisoniella, Sewardiella* and *Stephensoniella.*

Among the alpine liverworts of the Himalayas the best known are *Blepharostoma trichophyllum* Dum., *Lophozia alpestris* Evans., *Marchantia polymorpha* L., *Plagiochasma articulatum* Kash., *Preissia quadrata* Nees, *Riccia crystalina* - L., *Sauteria alpina* Nees, *S. spongiosa* Hatt. and *Scapania purpurea* Kash. Some of which reach an altitude of over 4,575 m. In the temperate Himalayas occur species of *Athalamia, Marchantia, Pellia, Riccardia, Fossombronia, Sewardiella, Madotheca, Lophocolea, Ruboulia* and *Solenostoma.* The subtropical and tropical belts of the Eastern Himalayas and other mountains are rich in species of *Calyogeia, Jungermannia, Plagiochasma, Plagiochila, Preissia, Mastigobryum, Lepidozia, Frullania, Lejeunea* and *Anthoceros.*

From the Nepal Himalayas are reported *Bryopteris trinitensis* L. et L., *Frullania wallichiana* Mitt., *Marchantia linearis* L. et L., *Plagiochasma cordatum* L. et L., *Madotheca revoluta* L. et L. and *Radula javanica* G.

Examples from Kumaun are *Anthoceros himalayesnsis* Kash., *Metzgeria furcata* Dum., *Notothylas levieri* Schiffn., *Plagiochasma appendiculatum* L. et L. and *P. articulatum* Kash.

Among the major forms in Assam mention may be made of *Anthoceros glandulosus* L. et L., *A. punctatus* L., *Frullania apiculata* Nees, *Plagiochasma articulatum* Kash., *Ptychanthus striatus* Nees and *Thysanthus spathulistipus* Lindb.

In the mountains of South India occur forms like *Madotheca nilgerriensis* Mont., *Dumortiera hirsuta* R. Bl. et Nees, *Frullania glomerata* L. et L., *Lunularia cruciata* Dum., *Marchantia nitida* L. et L., *Reboulia hemispherica* Raddi, *Schistochila aligera* st., *Strepsilejeunea neelgherriana* St., *Anthoceros erectus* Kash. and some 8 other species, *Mastigolejeunea repleta* St. and *Targionia hypophylla* L.

The epiphyllous liverworts of India are very imperfectly known. They occur on the leaf surface of trees mostly of wet evergreen forests of South India, Eastern Himalayas, Assam and the Andamans. About 27 species of these are so far known to occur in this country. Examples are *Leptocolea lanciloba* St., *Radula javanica* G., *Taeniolejeunea peraffinis* Zwickel, *Cololejeunea hispidissima* Herz. and *Leptolejeunea himalayensis* Pande et Misra.

Mitten's (1861) enumeration of hepatics included at least 39 genera with 205 species from India. Stephani (1885-1924) reported about 410 species from India. Kashyap, who was for many years leading botanist in India on liverworts, described several new genera and species and further additions were made by others. By 1952 (see Pande & Bharadwaj, 1952) the list of Indian liverworts rose to 550 species, and at present the figure is about 672. The Eastern Himalayas with about 330 representatives are the richest, South India has 225 species, and Western Himalayas have 170. The East Himalayan hepatic flora shows distinct affinities with that of the Malayan region and some members are common with China (12 species), Japan (26 species) and Australia (11 species). The Hepatic flora of Sikkim is remarkably similar to that of Yunnan and the Malayan region. South India has some species in common with Malaya and a few with China (7 species), Japan (23 species) and East Africa (17 species). The West Himalayan species show a greater affinity with Europe, and less with China (12 species) and Japan (13 species). The liverworts of the Western Ghats have affinities with those of Africa, and of the Eastern Ghats with those of Malaya, Java, Formosa, Sumatra and Borneo.

In the end, it must be said that many of our genera and species of liverworts are in need of critical study and monographic treatment.

Musci : The mosses flourish in a variety of habitats. On the dry faces of cliffs of gneiss and granite occur *Anoectangium walkeri* Broth., *Anomobryum cymbifolium* Broth., *Barbula comosa* Doz. et Molk., *Brachymenium walkeri* Broth., *Campylopus gracilis* Jaeg., *Hyophila cylindrica* Jaeg. and *Weisia edentula* Mitt. On pegmatite, lime and black loam occur *Buarbula indica* Brid., *Hyophila involuta* Jaeg. and *Rhodobryum giganteum* Par. A number of mosses, such as *Brachythecium procumbens* Jaeg., *Bryum wightii* Mitt., *Dicranella heteromalla* Schimp., *Fissidens lutescens* Broth., *Garckea phascoides* C. Müll. and *Trematodon ceylonensis* C. Mull., grow on the banks of streams and on clay in shady places. On comparatively dry banks we meet *Bryum ramosum* Mitt., *Campylopodium khasianu* Par., *Dicranella pomiformis* Jaeg. and *Funaria hygrometrica* Dill. var. *calvascens*.

Bryum apalodictyoides C. Mull., *Fissidens anomalus* Mont., *Leucobryum wightii* Mitt., *Leucoloma walkeri* Broth., *Tayloria schmidii* C. Müll. and *Trichosteleum monostictum* Broth., grow on dead and decayed tree trunks. On trees in dry open jungles we find

Brachymenium nepalense Hk., *Leucobryum hamillimum* Cardot, *Macromitrium leptocarpum* Broth., *Trachypus blandus* Mitt. and *T. crispatulus* Mitt. In very dense jungles, several mosses grow as felts on large trees and their branches and may loosely hang down like festoons. As examples may be mentioned *Aerobryum speciosum* Doz. et Molk., *Leucoloma renauldii* Broth., *Macromitrium moorcroftii* Schw., *M. sulcatum* Brid. and *Trichostomum hyalinoblastum* Broth. Some mosses form mats on tree trunks. *Brachymenium weisia* Hk. *Campylopus goughii* Jaeg., *Meteorium reclinatum* Mitt., and *Thamnium alopecurum* Bry. are examples of such mosses. A few species occur on plantation crops–*Papillaria fuscescens* Jaeg. on orange trees; and *Acrocryphaea concavifolia* Bry. and *Meteorium brevirameum* Broth. on coffee bushes. Species of *Sphagnum* are characteristic of bogs. *Fissidens grandifrons* Brid. grows on rocks directly under waterfalls.

Several species are known from the mountains. Of those growing between 1,200 m. and 1,800 m. the most important are *Bryum argenteum* L., *Mnium lycopodioides* Hk., *Philonotis falcta* Mitt. and *Tortula inermis* Mont. Between 1,800 m. and 2,700 m. grow *Bryum turbinatum* Hedw., *Desmatodon latifolium* Brid., *Funaria hygrometrica* Dill., *Grimmia leucophaea* Grew., *Plagiopus oederi* Limpr. and *Mnium medium* Br. et Sch. Among important species occurring between 2,700 m. and 3,300 m. are *Amphidium lapponicum* Schimp., *Barbula recurvifolia* Mitt., *Grimmia commutata* Hubn., *Trichosteleum brachypelma* Broth. and *Orthotrichum anomalum* Hedw. The high altitude (3,000 m.) mosses from Nepal Himalayas include *Bryum ventricosum* Dicks., *Dicranum himalayanum* Mitt., *Pleurozium schreberi* Mitt., and *Trachypodopsis crispatula* Fleisch. *Rhytidium rugosum* Kindb., grows at an altitude of 4,600 m. and species of *Andreaea* are seen at about 4,900 m.

Several species occurring in India are also met with in other parts of the world. Among others *Distichium inclinatum* Br. et Sch., *Fissidens grandifrons* Brid., *Grimmia ovata* Web. et Mohr. *Mnium lycopodioides* Hk. and *Weisia wimmeriana* Bry. occur in Europe. Species in common with North America are *Barbula vinealis* Brid., *Dicranum undulatum* Enrh., *Ditrichum tortile* Lindb., *Sphagnum teres* Angs. and *Timmiella anomala* Lmpr. *Campylopus comosus* Bry., *Dicranella heteromalla* Schimp. and *Herpetineurum toccoae* Card. are common with South America. The species in common with Africa are *Anoectangium euchloron* Mtt., *Ditrichum flexifolium* Hamp.,

Grimmia commutata Hube. and *Trichostomum cylidricum* C. Mull. and with Madagascar there are *Floribundaria floribunda* Fleisch. and *Philonotis laxissima* Mitt. The mosses in common with Australia and New Zealand are *Bartamia pomiformis* Hedw. and *Gymnostomum calcareum* Hornsch.

The following species occur in India as well as Indonesia, Malaysia part of Borneo and the Philippines : *Dicranella setifera* Jaeg., *Fissidens splachnobryoides* Broth.; *Leucobryum aduncum* Doz. et Molk., *Leucobryum javense* Mitt., *Microdus brasiliensis* Ther., *Philonotis falcata* Mitt., *P. fontana* Brid. and *Rhodobryum giganteum* Par. Species in common with China are *Cleistostoma ambigua* Mitt., *Dicranum perfalcatum* Broth., *Fissidens nobilis* Griff., *Macromitrium nepalense* Schw., *M. sulcatum* Brid., *Orthotrichum hookeri* Mitt., *Papillaria fuscenscens* Jaeg., *Ptychomitrium tortula* Mitt., *Rhacopilum schmidii* C. Müll., *Sphagnum junghuhnianum* Doz. et Molk., *Tortella fragilis* Limpr. and *Trachypodopsis crispatula* Fleisch. In common with Japan we have *Anisothecium rufescens* Lindb., *Brothera leana* C. Müll., *Leucobryum nilghiriense* C. Müll., *Leucoloma nitens* Par., *Sphagnum acutifolium* Enrh, and *Thysanomitrium blumei* Broth.

Out of 17 species of *Sphagnum* occurring in India, 7 are indigenous. The latter are as follows—*Sphagnum acutifolioides* Wans., *S. contortum* Wils., *S. cuspidatum* Enrh., *S. griffithianum* Warns, *S. khasianum* Mitt., *S. obtusifolium Griff.* and *S. ovatum* Hampe var. *cymbifolium. Bryum, Campylopus* and *Fissidens* are large genera, each with over 50 species. The Archidiaceae is a monogeneric family with the genus *Archidium* having 3 species in Kanara and the Palni hills. Similarly, the Encalyptaceae has only the genus *Encalypta* having 5 species in Punjab, Kashmir, North-western Himalayas and Ladakh.

Some new genera and species of mosses have been described from India. Among the former are *Dendrocyathophorum* and *Ortholobium,*. Of the several new species discovered in this country mention may be made of the following few : *Barbula dharwarensis* Dix., *Ctenidium stereodontoides* Dix., *Dicranum orthophylloides* Dix., *Duthiella mussooriensis* Reimers., *Fissidens karwarensis* Dix., *Macromitrium nilgirense* C. Müll., *Physcomirellopsis indica* Dix., and *Splachorbyum indicum* c. Müll. It must be noted that, so far, mosses have been collected and described only from certain restricted areas of this country. Extensive and systematic collections from all over the country and preparation of keys and checklists for easy identification of our common local mosses is long overdue.

Chapter 22
ALGAE

The algae constitute a large group of plants in which the plant body generally exhibits little or no differentiation of the vegetative organs into true stems, roots and leaves. They vary from simple unicellular forms, like *Euglena, Chlamydomonas, Protosiphon, Botrydium,* and the desmids and diatoms, to more complex multicellular forms. Many of the simpler forms are faultlessly symmetrical and are among the most beautiful objects in the world. In the largest forms the plant body may be differentiated into a blade, stipe (sometimes extremely long) and a specialized holdfast sometimes also accompanied by considerable internal differentiation. Particularly interesting are species of *Sargassum* for here the plant body is divisible into structures resembling leaves, stems, and branches. There are also bladders which look like tiny berries so that superficially the plants have the appearance of an aquatic angiosperm.

Algae occur in a variety of climatic and edaphic conditions in tropical, temperate, arctic, antarctic and alpine zone. This also occur in fresh water, brackish water, in marine situation, and on damp earth, walls, bricks, tree trunks, barks and leaves. Some forms are lithophytic, epiphytic, endophytic, symbiotic, parasitic or endozoic.

Some algae also occur at high altitudes. Among them are species of *Zygnema, Oedogonium* and *Vaucheria* together with some desmids and diatoms. *Hydrurus* and *Batrachospermum* occur as lithophytes up to 2,500 . At slightly lower elevations we have *Oocystis solitaria* Wittr., *Staurastrum cuspidatum* Breb. and a few species of *Cosmarium* and *Closterium*. Marine planktonic forms like *Hornellia marina* Subrahmanyam, *Trichodesmium erythraeum* Ehrenb. and some Dinophyceae are responsible for the yellow-green or green "discoloration", or "red tides", and phosphorescence of the sea.

Algae are classified into several phyla of which the more important are : *Cyanophyta* (blue-green algae), *Chlorophyta* (green algae), Charophyta (stoneworts), *Euglenophyta* (Euglenoids), Pyrrophyta (cryptomonads and dinoflagellates), *Chrysophyta* (yellow-green, golden-green algae and diatoms), *Phaeophyta* (brown algae), and *Rhodophyta* (red algae).

Cyanophyta : Members of the Cyanophyta are found in a great variety of situations exhibiting extremes of environmental and ecological conditions. In India there are more than 85 genera and 750 species belonging to 18 families. Of these the largest number of genera (15) fall under the Oscillatoriaceae. There is a new family Mastigocladopsidaceae of Indian origin with seven genera- *Camptylonemopsis, Iyengariella, Mastigocladopsis, Parsarthella, Spirulinopsis, Thackerella* and *Westiellopsis*.

Some interesting forms are known to inhabit the shells of molluscs, gastropods and barnacles. Some occur in dead corals. Some salt tolerant Myxophyceae of the genera *Anabaena, Anabaenoposis, Arthrospira, Myxosarcina* and *Synchococcus* form dense planktonic populations in the Sambhar lake in Rajasthan.

A few species of the Myxophyceae are known from hot springs at temperatures ranging from 30°C to 54°C. The more important among them are species of *Anabaena, Aphanocapsa, Aulosira, Gloeocaps, Oscillatoria, Phormidium, Plectonema* and *Scytonema*. Many of the blue-green algae occur in the soil. Among these the following are known to be nitrogenfixers : *Anabaena ambigua* Rao, *A. fertilissima* Rao, Aulosira *fertilissima* Ghosh, *Cylindrospermum sphaerica* Prasad, *Nostoc paludosum* Kutz. and *Tolypothrix tenuis* Jhos. Species of *Nostoc* and *Anabaena* live symbiotically in association with other plants such as *Anthoceros, Azolla* and *Cycas*. Some of the blue-green algae living in soil have been considered useful in reclaiming "usar" lands in Uttar Pradesh.

Several forms occur as plankton, sometimes in such abundance as to colour the entire body of water and form a fairly thick layer. The chief components of these waterblooms, mostly found only in alkaline waters, are *Microcystis, Arthrospira, spirulina, Anabaena, Oscillatoria* and *Trichodesmium*. When in great abundance, they not only harm fishes and animals but also offer serious problems in drinking water supplies by choking the sand filters.

Chlorophyta : Significant contributions have been made by Indian algologists on the green algae of paddy fields, hot springs, fresh and marine waters, and desiccated areas. Among new genera special mention may be made of *Characiosiphon, Chloranomala, Cylindrocapsopsis, Dendrocystis, Ecballocystopsis, Fritschiella, Gloeotilopsis, Heterotrichopsis, Hormidiella, Oocystaenium, Sirocladium,* and *Willeella*. Many botanists consider that land plant must have evolved from algae through a form like *Fritschiella*. *Characiosiphon* is also unique.

The Zygnemataceae and Desmidiaceae are large groups with a wide distribution. Among the latter, a new genus *Triplastrum* has been described from South India.

Species of *Chlamydomonas* and *Dunaliella* constitute major blooms in the Sambhar lake in brine of 17° to 26° Baumé, and *Dunaliella salina* Teod, grows in salt pans on the coast of Saurashtra in water having a salt concentration of 26° to 30° Baumé.

On an approximate estimate, at least 34 genera are marine : 9 are from the east coast and 28 genera from the west coast. The more well known of these are : *Acetabularia, Anadyomene, Avrainvillea, Boergesenia, Boodlea, Bryopsis, Caulerpa, Chaetomorpha, Chamaedoris, Cladophora, Codium, Dictyosphaeria, Enteromorpha, Halicystis, Halimeda, Microdictyon, Neomeris, Struvea, Trichosolen, Tydemania, Udotea, Ulva, Valonia,* and *Valoniopsis*. The total number of species is about 100. In inland fresh waters many of the Chlorophyta (especially desmids, Volvocales and Chlorococcales) form important constituents of the phytoplankton.

Charophyta : The members of this group form extensive underwater meadows in ponds and rivers and have structures which bear a superficial resemblance to roots, stems and leaves. Until 1822 only two genera of the Charophyta were known from India : *Chara* and *Nitella*. *Nitellopsis, Lychnothamnus* and *Tolypella* were reported later. Thus, the group is now represented in India by *Chara (26 species),* Nitella (34 species), *Lychnothamnus* (1 species), *Nitellopsis* (1 species), and *Tolypella* (3 species). Sixteen species of *Chara* and one of *Nitellites* are represented in the Deccan Intertrappean fossil beds.

Chrysophyta : The genus *Vaucheria* is represented by 11 species. Of these *V. hamata* Walz. is very common on the soil in the spring season. *V. terrestris* Lyngbye em Walz. occurs in Kashmir in the Amarnath cave at 3,885 m. and *V. sessilis* (Vanch.) Dc. in the cave-like vaults on the sides of the Verinag spring. *Botrydium* is characteristic of drying mud and is represented by *B. granulatum* Grev., *B. tuberosum* Iyen. and *B. divisum* Iyn *Hydrurus, Synura, Dinobryon, Mallomonas* and *Chrysodictyon* are the chief members of the Chrysophyceae. Of these *Hydrurus* is found only in the cool waters of mountain streams as in Kashmir.

The diatom flora of inland as well as marine waters has been studied by several botanists. A large number of new species have been recorded for which reference must be made to other sources.

Euglenophyta : The Euglenophytes are all inhabitant of foul water. There are 11 freshwater genera and 1 marine genus *Protoeuglena.* An interesting report is that of *Cladospongia* which is a colourless form belonging to the order Protomastigineane. One or two species of *Euglena* are frequently responsible for the red scums of stagnant inland waters.

Phaeophyta : Apart from *Sargassum,* there are at least 30 other genera with 75 species. All of these are marine and include *Chnoospora, Cladostephus, Colpomenia, Cystophyllum, Dictyota, Dictyopteris, Ectocarpus, Hecatonema, Hormophysa, Hydroclathrus, Mesogloea, Nemacystus, Padina, Spathoglossum, Sphcelaria, Streblonema, Turbinaria* and *Zonaria. Iyengaria* is a new genus with the species *I. stellata* Boerg.

Rhodophyta : No less than 125 genera with about 300 species occur in India. The chief of these are; *Acanthophora, Agardhiella, Amphiroa, Antithamnion, Asparagopsis, Batrachospermum, Botryocladia, Caloglossa, Ceramium, Champia, Chondria, Chrysymenia, Claudea, Coelarthrum, Corallina, Dasya, Dictyurus, Enantiocladia, Galaxaura, Gastroclonium, Gelidiella, Gelidium, Gigartina, Gracilaria, Grateloupia, Gymnothamnion, Halymenia, Helminthocladia,Heterosiphonia, Hypnea,*

Hypoglossum, Laurencia, Lemanea, Liagora, Neurymenia, Nitophyllum, Polysiphonia, Porphyra, Rhodymenia, Sarcodia, Sarcomenia, Sarconema, Scinaia, Spyridia, Thorea and *Vanvoorstia.* A new family Corynomorphaceae has been created for the genus *Corynormopha.*

While most of the red algae are marine, a few freshwater representatives are also known. Among the latter are *Acrochaetium, Batrachospermum, Compsopogon, Sirodotia* and *Thorea, Compsopogon* is represented by three species, *C. coeruleus* Mont., *C. hookeri* Mont. and *C. iyengarii* Krish.

Chapter 23

Ecology and Distribution of the Marine Forms

There are several places, both on the Peninsular coast and in the nearby islands and archipelagos, which show a rich algal vegetation. Among these are : Dwarka, Mul Dwarka and Okha in the Kutch-Saurashtra area; Bombay, Malvan and Karwar further down the Western Coast; Kannyakumari at the southernmost extremity; Rameswaram Pamban, Kursadi. Shingla and neighbouring islands. Tuticorin, Here Island and Church Island, all in the southern part of the eastern coast; and Mahabalipuram, Madras and Waltair further up on the eastern coast. In addition, the Laccadives, the Andamans and the Nicobars also provide rich algal collections.

In the various islands in the Gulf of Manaar and Laccadives, the substratum is mostly coralline conglomerate or sandstone. In the Andamans and Nicobars coralline and other rocky substrata are met with. In the other localities, particularly on the Peninsular coast, rocks of laterite, sandstone, altered traps, gneiss or schists serve as substrata. Certain localities like Long Island and Port Blair in the Andamans, Kursadi. Hare Island, Shingle Island and Church Island on the eastern peninsular coast, and Okha-Dwarka on the western coast are particularly favourable for algal collections as a large stretch of the coast is uncovered at low tide.

In all areas favourable for luxuriant algal growth, species belonging to the Chlorophyta, Phaeophyta and Rhodophyta are well represented. Although it is often difficult to demarcate clearcut zones due to the complexity of the associations, it is nevertheless possible in most areas to divide the algal vegetation into some broad belts, viz. the infra-littoral belt, extending from low watermark to deeper water; the littoral belt, extending between high and low water marks and which may often show three clear zones (the lower littoral, the mid-littoral and the upper littoral); and the supralittoral belt lying above high watermark.

Of the red algae, *Chrysymenia uvaria* J.Ag., *Halymenia dilatata* Zan., *Dictyrus purpurascens Bory, Neurymenia fraxinifolia* J.Ag., *Scinaia carnosa* Harv. are some of the typical forms met with in the infra-littoral region. Others like *Halymenia venusta* Boregs., *Asparagopsis taxiformis* Coll. et Herv. and *Botryocladia leptopoda* Kylin, though characteristic of the infra-littoral belt, are sometimes found further up in the littoral region. The lower littoral zone is populated mostly by species of *Gelidium, Polysiphonia, Ceramium, Laurencia* and *Gracilaria,* while the midlittoral and upper littoral regions show more of brown and green algal communities and comparatively fewer red algae. *Colpomenia sinuosa* Derb. et Sol., *Iyengaria stellata* Boergs. and *Hydroclathrus clathratus* Howe and species of *Padina, Sargassum* and *Dictyota* are some of the typical brown algae of this region. The green algae are represented by *Codium, Caulerpa, Ulva, Enteromorpha, Chaetomorpha,* and *Cladophora.* Within the tidal limits, there are rock pools of different sizes and depths which harbour interesting algal associations. In the very deep rock pools are seen red algae like *Botryocaladia leptopoda* Kylin and *Galaxura oblongata* Lamx. intermingled with brown and green algae. In pools higher up we see the red algae-like species of *Amphiroa, Jania,* and *Gracilaria* along with *Champia parvula* Hary., *Heterosiphonia muelleri* De Toni, and *Liagora ceranoides* Lamour; species of *Padina. Sargassum* and *Dictyota* and other brown algae; and green algae like *Acetabularia mobii* Solms-Labauch, *Chamaedoris auriculata* Boergs. and species of *Enteromorpha, Ulva, Caulerpa,* and *Chaetomorpha.* In the supra-littoral region the rocks, exposed to heavy swell and suf, harbour a characteristic flora comprising species of *Chaetomorpha, Cladophora* and *Halimeda* of the green algae; *Ectocarpus, Myriogloea, Namacystus,* and *Chnoospora,* of the brown algae; *Porphyra, Liagora, Sarcodia, Grateloupia, Gracilaria* of the red algae; and blue-green algae like species of *Lyngbya. Microcoleus* and *Calothrix. Boergesenia, Udotea* and at places *Avrainvillea* are found on reefs and coralline substrata heavily silted with fine sand and mud to form flats which frequently get exposed at low tide. In such situations *Acetabularia* and *Neomeris* are also found at times growing on fragments of dead corals. In shallow lagoons, several brown algae like *Cystophyllum muricatum* J.Ag., *Dictyopteris delicatula* Lamour., *Hormophysa triquetra* Kutz, *Spathoglossum asperum* J.Ag., *Stoechospermum marginale* Kutz, *Spathoglossum asperum* J.Ag., *Stoechospermum marginale* Kutz, and species of *Turbinaria, Sargassum* and *Padina* occur associated with red algae like *Dictyurus, Acanthophora, Gracilaria* and *Hypnea.*

The brackish water mangrove swamps and salt marshes which occur scattered all along the coast also show very characteristic algal communities in which all the major groups are represented.

From the available information on the marine algal flora of Indian coasts it appears that there is considerable affinity with the flora of Mauritius (35.8% of the species are common to both floras) and the Atlantic coasts of the USA and Europe (22.7% of the species are common to both floras). About 22.5% of the species are common to Indonesia and India, and 22% to the West Indies and India. In the Pacific zone, Australia claims more species in common with India (20.3%) than Japan (20.1%). Thus, the Indian coast harbours a complex variety of algal vegetation, comprising many species which have extensive distribution both in tropical and temperate seas, and a few even extending to the Arctic seas. In comparison with the Peninsular flora, those of the Laccadives and Andamans have a number of species peculiar to each and not in common with the Indian coastal flora although some of these species have a very wide geographical distribution.

Economic Importance : Several marine forms and a few freshwater ones are edible. Special mention may be made of *Caulerpa, Enteromorpha, Gracilaria, Hydroclathrus* and *Ulva*. *Ulva fasciata* Delile is found in quantity on the Gujarat coast and has a protein content of dry seaweed of 31%. There is, therefore, the possibility of cultivating this species for food purpose in sea water.

Among the genera that can be used as sources of agar the most important are *Gelidium, Gracilaria, Hypnea* and *Sarconema.* The Indian agar potential is estimated at 13 metric tonnes annual, with an average yield of 28% on dry seaweed of the different species of agarophytes, while the Indian consumption is about 30 metric tonnes annually.

Chapter 24
FUNGI

The systematic study of fungi in India began only in the last quarter of the 19th century. Until about 1875 all collections of Indian fungi were being sent for study and identification to Europe and for many years M.J. Berkeley was the chief determiner of Indian fungi. From 1875 onward such work began to be undertaken in India itself. D.D. Cunningham and A. Barclay were the pioneers in this field, and their studies on the Mucorales, the Ustilaginales and the Himalayan rusts are still looked upon with high regard, K.R. Kirtikar studies the agarics and the Gasteromycetes. The arrival of E.J. Butler in India at the turn of the 20th century gave a special impetus to the subject. Before his return to the U.K. in 1920, his all time classic—*Fungi and Disease in Plants*-had already been published. He also laid the foundation of the *Herbarium Cryptogamiae Indiae Orientalis,* which is now our national herbarium for fungi and is located in the Indian Agricultural Research Institute in New Delhi. During the last 40 years Indian mycologists have taken an increasing share in a taxonomic study of fungi and among them special mention may be made of the following; J.F. Dastur, J.H. Mitter, B.B. Mundkur, C.V. Subramanian, and M.J. Thirumalachar. The year 1931 saw the publication of the first all India list of fungi–*The Fungi of India* – by E.J. Butler and G.R. Bisby. Recently it has gone through a new edition and is now a valuable work of reference.

Due to the diversity of the Indian climate we have many representatives of the fungal flora of Europe as well as of the Tropics. However, in comparison with the phanerogams, the fungi are still only partially explored and the total number so far known is just about five thousand.

The Myxomycetes or slime fungi have received a fair amount of attention in recent times and over 205 species are now recorded. The commones genera are *Arcyria, Badhamia, Ceratiomyxa, Comatricha, Craterium, Cribraria, Diachea, Dictydium, Diderma, Didymium, Fuligo, Hemitrichia, Lamproderma, Lycogala, Perichaena, Physarella, Physarum, Reticularia, Stemonitis* and *Trichia.*

Of the Phycomycetes there are 324 species included under 64 genera. Among the lower Phycomycetes the genus *Synchytrium* is represented by more than 58 species and the genus *Physoderma* by 18 species. The Blastocladiales have 4 species under *Allomyces* and 4 under *Blastocladia.* Among the Peronosporales, the chief genera–*Albugo, Bremia, Peronospora, Phytophthora, Plasmopara, Pythium* and *Sclerospora* — are each represented by several species. Fifteen genera of the Mucorales are known of which the commonest are *Absidia, Choanephora, Cunninghamella Mucor, Pllobolus, Rhizopus* and *Syncephalis.* There are two genera representing the Entomophthorales : *Conidlobolus* and *Entomophthora.*

The Hemiascomycetes represented by *Eremascus, Hansenula, Nematospora, Saccharomyces, Taphrina, Protomyces* and *Protomycopsis.* Most of the apothecial fungi are represented by a single species except genera like *Ascobolus, Dasyscypha, Humaria, Morchella, Peziza* and *Pseudopeziza, Morchella esculenta* Pers, grows in Kashmir, parts of Punjab and Naini Tal and is eaten in North India as a delicacy. The Sphaeriales are represented by a large number of genera of which *Chaetomium, Glomerella, Mycosphaerella, Physalospora, Rosellinia* and *Xylaria* are the most common. *Phyllachora* has as many as 62 species. *Claviceps purpurea* Tul., the ergot fungus, is cultivated on rye in the Nilgiris and Darjeeling. Most of the genera of the Erysiphaceae are present, although their perfect stages are not always seen except in the cooler areas of North India. There are no less than 58 species of *Meliola.*

The Ustilaginales and the Uredinales are very well represented. *Ustilago* has 56 species; *Sphacelotheca* has 53; and *Urocystis, Tilletia, Entyloma* and *Sorosporium* have also several species each. *Puccinia* is represented by 234 species, *Uromyces* by 83, *Ravenelia* by 28, *Melampsora* by 14, *Phakopsora* by 15 and *Hemileia* by 11 species. Among the important genera of the Agaricales may be mentioned *Agaricus, Armillaria, Boletus, Collybia, Daedalea, Fomes, Ganoderma, Hydnum, Lentinus, Lenzites, Lepiota, Polyporus, Polystictus* and *Poria* whereas *Lycoperdon, Podaxis, Scleroderma* and *Tylostoma* represent the Gasteromycetes. Many of the agarics are edible and are widely distributed in India. *Cantharellus cibarius* Fr. is particularly common in the hills of the northern and eastern parts. It grows during the rainy season and is regarded as a delicacy. Another edible mushroom *Volvaria diplasia* Sacc., occurs in Bengal and Madras. *Phellorinia inquinans* Berk. and *Calvatia gigantea* Fr. (Gasteromycetes) are found in the plains of North India. The latter looks like a pumpkin and is eaten only in the young stage when the inner portion is still white.

Among the Deuteromycetes the Moniliales have the largest number of species. *Cercospora* has 270 species while *Helminthosporium, Fusarium* and *Alternaria* are also large genera. Common representatives of the other orders are *Ascochyta, Colletotrichum, Diplodia, Macrophoma, Pestalotiopsis, Phoma, Phyllosticta* and *Septoria.*

During recent years several new genera have been discovered from different parts of the country. Among these are *Saksenaea* (Mucorales), *Bagcheea* (Sphaeriales), *Mundkurella, Zundelula Narasimhania* (Ustilaginales), *Dasturella, Ceropsora, Kernella Didymopsorella* (Uredinales), *Alpakesa, Bahusandhika, Dwayabeeja, Pseudotorula* and *Anthasthoopa* (Deuteromycetes).

Plant Diseases caused by Fungi

Of the fungal diseases the most important is the wheat rust which is estimated to cause an annual loss of about 50 lakh rupees. The Bengal famine of 1942 has been partly attributed to the helminthosporium disease of rice, while the red rot of sugar-cane has been responsible for several epidemics from time to time.

The club root of crucifers, caused by *Plasmodiophcra brassicae* Wor., is common in the hills of South India. The "damping off" of tobacco, tomato, and crucifers, caused by *Pythium debaryanum* Hesse and *P. aphanidermatum* Firzp., is widespread. These two fungi also cause the fruit rot of cucurbits, foot rot of papaya and soft rot of ginger. The late blight of potato, caused by *Phytophthora infestans* de Bary, occurs regularly in the hills and occasionally in the plains when the weather conditions are moderate. *P. palmivora* Butler causing the bud rot of palms causes much damage on the east coast of South India. It is also responsible for the fruit rot and nut fall (Koleroga) of areca in Western Peninsular India and Assam. *Peronospora* causing the downy mildew of several winter crops and *Albugo* causing the white rust of crucifers are common but the diseases are not serious. *Sclerospora graminicola* Schrot., is responsible for the green ear disease of *bajra* which is quite damaging at times.

The stem gall of coriander (*Protomyces macrosporus* Ung.) is widespread and sometimes affects seed setting in this crop. *Taphrina deformans* Tul. causes the leaf curl of peaches in Kashmir, Kulu, Kumaun, Simla, Nilgiri and Palni hills. *T. maculans,* Butler, causing the leaf-spot of turmeric, is very troublesome in Gujarāt, the Northern Circars, Orissa and Andhra Pradesh. *Aspergillus niger* Van Tiegh. causes a soft rot of apples in Uttar Pradesh and seedling blight of ground-nut in North India. *Erysiphe polygoni* DC. is a common mildew affecting the pea, lentil, and other leguminous crops. *E. graminis* DC., the powdery mildew of cereals, does much harm to wheat and barley in the hills and submontane districts. The powdery mildew of vines is also common but not in an epidemic form. *Glomerella tucumanesis* Arx et Müll. causes the red rot of sugar-cane which is very serious in North and East India. Gram blight, caused by *Mycosphaerella rabiei* Kov. is damaging in North India. *Helminthosporium oryzae* Breda de Haan is widespread on rice in Assam, Bengal and parts of Madras.

There are smuts affecting all the important graminaceous crops. *Ustilago tritici* Jens. causing the loose smut of wheat, is common in most tracts. *Tilletia indica* Mitra (Karnal bunt) causes a partial bunt of wheat in the submontane regions of Punjab and Uttar Pradesh. The flag smut of wheat occurs in the Punjab and the root smut gall of mustard is found in Bihar. Both the diseases are caused by *Urocystis.* All the three rusts of wheat are widespread but the laternate hosts do not play any role in their perpetuation. The rusts of linseed, coffee and gram are common. The conifers are also subject to a number of rusts. *Cronartium himalayense* Bagchee is endemic and parasites *Pinus roxburghii* Sarg. The blister blight of tea (causal organism – *Exobasidium vexans* Massee) is prevalent in a serious form in Darjeeling and Madras. *Pellicularia salmonicolor* Dast. causes the pink disease of orange in Andhra Pradesh. It is widespread among other plantation crops such as tea, rubber, coffee, cinchona and mango. *Ganoderma lucidum* Karst., causing spongy rot, has been recorded from various parts of India on *Casuarina equisetifolia* Forst., *Areca catechu* L., *Pongamia pinnata* Pierre, *Guazuma tomentosa* H.B.K., *Acacia* spp, *Pterocarpus marsupium* Roxb., *Cocos nucifera* L. and several others. *Trametes pini* Lloyd causes the red ring rot of many Himalayan conifers. Other wood rotting fungi of India are *Armillaria mellea* Quel., *Fomes badius* Berk., *F. senex* Nes et Mont., *Polyporus gilvus* Schwein., and *P. palustris* Berk. et Curt.

Among the Deuteromycetes the most important fungal pathogens are *Fusarium* and *Piricularia. P. oryzae* Cavara causes the blast of paddy which is especially destructive in Madras, the losses being 30-35% in certain localities. *Fusarium udum* Butler causing the wilt of pigeo peas, does much harm in parts of Maharashtra, Madhya Pradesh, Uttar Pradesh and Bihar, *F. vasinfectum* Atkins. is prevalent in the black cotton soil area of the Deccan and Gujarat. The stem rot of jute caused by *Macrophomina phaseoli* Ashby is a limiting factor in the successful cultivation of this valuable crop.

Till 1931, only 2,351 species of fungi were known with 75 species under the Phycomycetes, 476 under the Ascomycetes, 1,339 under the Basidiomycetes, and 461 under the Fungi Imperfecti. By 1938 this number increased to 2,868, and in 1951 to 3,680. During the last 11 years, 64 new genera and 1,150 new species have been recorded—a good indication of the activity of mycologists and plant pathologists throughout the country.

Chapter 25

LICHENS

Although lichens occur in abundance in the mountains of India, their study has so far received only scant attention. C. Montagne, C. Babington, W. Nylander, J. Muller Argoviensis, J. Stirton, A. Jatta, R. Paulson and A.L. Smith made the initial contributions to the lichen flora of India. G.L. Chopra (1934) gave a comprehensive and illustrated account of 75 lichens collected from Darjeeling and the Sikkim Himalayas. Presently, D.D. Awasthi at Lucknow has taken much interest in a study of the Indian lichens.

The chief interest of lichens lies in their comprising two distinct symbiotic partners, a fungus and an algae. The algal cells are enveloped in the intricate felted mass of fungal hyphae, and the two members mutually benefit by this association. The fungal element generally belongs to the Ascomycetes (only three genera are basidiomycetous and of these only one is known from India), the algal partner being a member of the Myxophyceae or the Chlorophyceae. The lichens form incrustations (crustose forms), foliaceous masses (foliose forms), or branching forms (Fruticose forms) on rocks, soil, tree-trunks, leaves and other suitable substrata.

Over 800 species are known from the Indian subcontinent (including Nepal) and the Andamans. Of these the following families are represented by a large number of species : Usneaceae

(78 spp.), Parmeliaceae (74 spp.), Graphidaceae (71 spp.), Lecanoraceae (62 spp.), Physciaceae (61 spp.), Lecideaceae (60 spp.), Cladoniaceae (53 spp.), and Pyrenulaceae (45 spp.) The other families which have a smaller number of species are Collemaceae (28 spp.), Stictaceae (18 spp.), and Peltigeraceae (1 spp.) While families like the Dermatocarpaceae, Callciaceae, Sphaerophoraceae. Diploschistaceae, and Teloschistaceae are represented by less than 5 species each.

The crustose lichens occur abundantly in India in deciduous and mangrove forests and alpine regions on tree and rocky surfaces. Only the temperate regions are really favourable for the best growth of lichens, and some species reach up to an altitude of 5,000 m. A few also inhabit snow and glacial beds, growing as lithophilous forms on rocks and can withstand a burial under the snow for the greater part of the year. *Dermatocarpon miniatum* Th.-Fr. grows on exposed rocks, and *Caloplaca murorum* Th.-Fr. on volcanic rocks. *Endocarpon pusilum* Hedw., *Peltigera malacea* Ach. and *Cladonia rangiformis* Hoffm. grow on hard and dry soil. *Rinodina sophodes* Th.-Fr. occurs in salt marshes and *Cetrarai ambigua* Bab. on hard soil in alpine bets. A good number of species such as *Lobaria pulmonaria* Hoffm., *Peltigera polydactyla* Hoff. and *Sticta fuliginosa* Ach. grow on the barks of various trees along with mosses. Altitudinally, *Coccocarpia pellita* Mull-Arg. and *Sticta weigeli* Wain. are the common forms between 1,220 m. and 1,525 m.; *Leptogium saturninun* Nyl. at 1,830 m.; *L. menziesii* Mont., *Lobaria pulmonaria* Hoffm. and *Peltigera canica* Willd. between 2,130 m. and 2,440 m.; *Cladonia furcata* Schrad. at about 2,740 m.; *Parmelia conspersa* Ach. at 3,655 m., *Lobaria retigera Trevis*, between 4,000 m. and 4,500 m.; and *Stereocaulon tomentosum* Fr. at 5,000 m.

A few species are of economic importance. *Lobaria pulmonaria* Hoffm. is used for asthma and lung troubles, *Parmelia perforata* Ach. as a diuretic, *P. saxatilis* Ach. for epilepsy and *Peltigera canina* Willd. for hydrophobia and jaundice. Several species yield dyes : *Diploschistes scruposus* Norm., *Parmelia physodes* Ach. (a brown dye), *Gyrophora lecanocarpoides* Th.-Fr. *Umbilicaria pustulata* Tuck. (a red brown dye) and *Parmelia olivacea* Nyl. (a yellow dye). *Parmelia physodes* Ach. and *Ramalina fraxinea* Ach. are considered suitable substitutes for gum-arabie, and *Ramalina farinacea* Ach. and *R. fraxinea* Ach. are used for making cosmetics. *Lobaria pulmonaria*

Hoffm. is used in tanning and brewery, and *Dermatocarpon moulinsii* A. Zahlbr. is a substitute for cork for lining entomological collecting boxes. *Cladonia alpestris* Rabh. which occurs at higher elevations of the Himalayas, has been used in the Arctic areas as an article of food for the reindeer.

Chapter 26

Botanical Regions of India and their Floristic Compositions

In their *Introductory Essay to the Flora Indica,* Hooker and Thomson (1885) attempted a phytogeographical analysis of the Indian flora as a whole; this was later incorporated by Hooker in *The Imperial Gazetteer of India* (1907).

Hooker divided the then British possessions of India into nine botanical regions, using the number of species of the ten largest families in each region as the most important criterion for his classification. The nine regions are : (1) the Eastern Himalayas extending from Sikkim to the Mishmi hills in Upper Assam; (2) the Western Himalayas extending from Kumaun to Chitral; (3) the Indus plain including the Punjab, Sind and Rajasthan west of Aravalli range and Yamuna river; Kutch; and Northern Gujarat; (4) the Ganga plain, from the Aravalli hills and Yamuna river to Bengal; the Sundarbans; the plains of Assam and the low country of Orissa north of the Mahanadi river; (5) Malabar, the humid belt of hilly or mountainous country extending along the western side of the Peninsula from Southern Gujarat to Cape Comorin, including Southern Gujarat, the southern half of Kathiawar, the Konkan, Kanara, Kerala and the Laccadive Islands, (6) the Deccan, the comparatively dry elevated tableland of the Peninsula east of Malabar and south of the Ganga and Indus plains, together with, as a subregion, the lowlying strip of coastland extending from Orissa to the Tirunelveli Distric known as the Coromandel Coast;

(7) Ceylon and the Maldives Islands; (8) Burma; and (9) the Malay Peninsula. Hooker was not sure where to place the Andaman and Nicobar Islands.

In the delimitation of these areas, it is difficult sometimes to apportion large areas into one or the other of two contiguous botanical regions. This is to be expected, since the changes in environmental conditions are gradual and not sudden. For instance, the north-western half of Kathiawar is botanically similar to Sind, and the south-eastern to the Konkan. Nor is it possible to draw an absolute boundary line between the floras of the Indus and of the Ganga plains. A number of Upper Ganga plants intrude into the Indus plain and those of the Rajasthan desert into the Ganga plain. Again, the eastern limit of the Malabar region is undefinable, because of the number of spurs and valley from its hills which project far into the Deccan region, almost crossing it. They carry with them types of the Malabar flora, which towards, its northern limit, mingle with the floras of the Deccan and of the Indus and the Ganga plains.

Calder (1937) recognized only six main divisions of India : (1) the North-western Himalayas, (2) the Eastern Himalayas, (3) the Indus plain, (4) the Ganga plain, (5) the Deccan (with one eastern subprovince), and (6) Malabar.

Chatterjee (1939) based his divisions mainly on the endemic content of the dicotylendons in the different areas. He excluded Ceylon, the Maldives and Malaysia since these have floras distinctly different from the flora of India. Assam, which was included by Hooker in the Ganga plain region, was considered separate region because of its distinctive flora (*cf.* Clarke, 1898). The Himalayas have been divided into three instead of two regions. Further, in consideration of the definitely older geological age of Peninsular India, the Deccan and Malabar regions have been regarded as floristically older. The revised floristic regions of Chatterjee, as applicable to present-day India (excluding Nepal, Pakistan and Burma), are as follows : (1) Deccan, (2) Malabar, (3) Indus plain, (4) Ganga plain, (5) Assam, (6) Eastern Himalayas and (7) Western Himalayas. The Andamans, may be taken as the eighth botanical region of India.

From point of view of their relatively low rainfall and humidity, the Deccan, the Indus plain, and the Western Himalayas show a marked contrast with Malabar, Lower Ganga plain, Assam and the Eastern Himalayas. The striking floristic differences between these two groups of regions are, therefore, quite understandable. Altitude is the chief factor in the characterization of the mountain vegetation of India, particularly in the Himalayas, Soil is a factor of more local significance. The members of the Dipterocarpaceae can be cited as a good example of preferential distribution with reference to rainfall and soil. In *Dipterocapus,* there are species which favour a drier environment, such as *D. obtusifolia* Teysm. and *D. tuberculatus* Roxb., and others which are of a more hygrophilous type such as *D. turbinatus* Gaertn. f., *D. indicus* Bedd., *D. pilosus* Roxb. and *D. alatus* Roxb. In general, these two groups show further contrast in that the xerophilous species almost always occur gregariously and are deciduous, while the hygrophilous species occur sporadically and are evergreen.

Peninsular India, one of the three regions into which India is divisible on the basis of the percentage of endemic species, floristically comprises the deccan and the Malabar regions, both of which together have a high endemic content next only to the Himalayas.

The Deccan Region : This comprises the entire, comparatively dry elevated tableland of the Peninsula east of Malabar and south of the Indo-Ganga plain. The hills of the Vindhyas and Eastern Ghats fall in this region. The Coromandel Coast extending from Orissa to Tirunelveli may be considered as a subregion. Over the greater part of the Deccan region, the rainfall is less than 100 cm., and this amount is exceeded only in certain elevated parts which intercept the monsoon currents. However, the Coromandel subregion (also termed Carnatic subregion) receives the full benefit of the north-east monsoon and has a rainfall ranging from 62.5 to 162.5 cm.

Among the palms of the Deccam region are *Phoenix sylvestris* Roxb., *P. robusta* Hook., f., *P. acaulis* Roxb., *P. humilis* Royle, *Calamus viminalis* Wild., *C. pseudotenuis* Becc., *C. rotang* L. and *Borassum flabellifer* L.

Successful *Casuarina* plantations have been raised along the coast on sandy soil.

Malabar Region : This comprises the excessively humid (rainfall, more than 200 cm.) belt of mountain country running parallel to the west coast of the Peninsula. It is mostly a hilly country, and except in the north, the mountains often rise abruptly from the flat coast of the Arabian Sea. Its abrupt western face is clothed wth a luxuriant, evergreen forest merging towards the drier north into the elements of the Deccan and the Indus plain floras. The eastern face slopes gradually into the elevated plateau of the Deccan but there are many spurs projecting far into the Deccan region, often enclosing vallyes with a Malabar flora. A particularly great break occurs in the chain at the latitude of 11° N where a transverse valley separates Southern Kerala from the mountains north of it and carries species characteristic of the Malabar flora almost across the Peninsula. To this region belong the Nilgiri hills.

The endemic palms of this region are : *Pinanga dicksonii* Bl., *Bentinckia coddapanna* Berry, *Calamus rheedei* Griff., *C. huegelianus* Mart., *C. brandisii* Becc., and *C. gamblei* Becc. Among other wild palms are species of *Phoenix, Caryota, Calamus* and *Corypha.* Of the commercial crops the most important are betelnut *Areca catechu* L., coconut (*Cocos nucifera* L.), palmyra (*Borassus flabellifer* L.; pepper (*Piper nigrum* L.), coffee (*Coffea* spp.) and tea (*Camellia sinensis* O. Ktze). In more recent times *Hevea brasiliensis* Mull.-Arg. (rubber), *Anacardium occidentale* L. cashew-nut) and *Eucalyptus* spp. have been introduced successfully in suitable areas of this region, rubber in the very humid regions, cashew-nut along the coast, and eucalyptus in the Nilgiri and other hills. The coconut forms a major element in the economy of the Kerala State and it is common to see this palm lining the lagoons and canals of the coastline.

The Indo-Ganga plain, which is the poorest in endemic content, is divisible into the Indus plain region and the Ganga plain region.

Indus Plain Region : This region comprises the plains of Punjab, Rajasthan west of the Aravalli range and Yamuna river, Kutch and Northern Gujarat. The rainfall is less than 75 cm. generally and in the driest regions of the desert area less than 12.5 cm. in the year.

The only indigenous palms in the Indus plain region are *Phoenix sylvestris* Roxb. and *Nannorrhops ritchieana* H. Wendl. The latter finds its north-eastern limit in the Salt Range and the south-western limit in Sind and Baluchistan, both of which areas are now in Pakistan.

Prosopis juliflora DC. and *P. glandulosa* Torr. of the arid regions of Mexico and Central America have been successfully introduced in connection with the soil conservation and afforestation of dry and desert areas.

Ganga Plain Region : This region stretches from the Aravalli hills and the Yamuna river eastward to Bengal, including the Sundarbans and the low country of Orissa north of the Mahanadi river. The bulk of this tract has been under cultivation from very early times. The forests, where they exist, are of widely different types. The sal forests of Oudh are probably mere remnants of the great sub-Himalayan sal belt, which at one time covered a much larger area than now and stretched for some distance into the adjoining plains; the lower Ganga subregion, comprising both Bihar and Bengal, is much more humid than the upper region. *Areca, Phoenix, Borassus* and *Cocos* are cultivated; of indigenous palms mention may be made of *Corypha, Calamus* and *Daemonorops.* The third subregion comprises the Sundarbans, which borders on the Bay of Bengal. The chief plants here are : *Heritiera fomes* Buch.-Ham. (Sundri), *Excoecaria agallocha* L., *Sonneratia apetala* Buch.-Ham. *S. caseolaris* Engl., *Xylocarpus molluccensis* Roem., *X. granatum* Koen., *Amoora cucullata* Roxb., *Aegiceras corniculatum* Blanco.*Cynometra mimosoides* wall., *Avicennia officinalis* L. and the mangroves *Ceriops tagal* C.B. Robbins. *C. roxburghiana* Arn., *Kandella rheedii* W. et A., *Rhizophora mucronata* Lam. and *Bruguiera conjugata* Merr. The palm *Nipa fruitcans* Wurmb. is gregarious in the swamps and on river banks, and *Phoenix paludosa* Roxb. is found in drier localities. A species of *Calamus* and another of *Daemonorops* are common.

Assam Region : This comprises the Brahmaputra and Surma valleys together with the intervening hill ranges-the Garo, Khasi and Jaintia hills, and also the Nowgong, Naga, Patkal, Manipur and Lushai hills on the eastern and south-eastern frontiers Assam. Over the greater part of this region the rainfall exceeds 200 cm., while Cherrapunji in the Khasi hills, with a normal rainfall of 1,080 cm., is reputedly the rainiest spot in the world. The vegetation is luxuriant and the valleys, where they are not under tea or agricultural crops, are clothed with expanses of tall savannah grass or with dense forest often of an evergreen type.

The hill forests of the Assam region approximate in type to those of the Eastern Himalayan region, except that there is no alpine zone. These hill forests may be separated broadly into evergreen forests, broad-leaved forests and pine forests.

Shifting cultivation has destroyed much of the natural forests growth on these hills. The hill-tops of the Assam region, like those of the Nilgiris, are open grasslands with trees and shrubs identical with or closely related to those of the Nilgiris. The mountains to the east are often covered with bamboos.

Among palms of narrow distribution in the Assam region are : *Areca nagensis* Griff., *Pinanga griffithii* Becc., *P. hookeriana* Becc., *Didymosperma nana* Wendl., *D. gracilis* Hook. f., and *Plectocomia khasyana* Griff. Besides these, there are a number of others of wider distribution, like *Wallichia densiflora* Mart., *Caryota* spp., *Licuala peltata* Roxb., *Phoenix* spp., *Daemonorops* sp., *Zalacca* sp., and several species of *Calamus*.

Eastern Himalayan Region : This region, extending from Sikkim eastwards, embraces the most humid portion of the Himalayan range. Darjeeling, Kurseong and other places are located in this part of the Himalayas. The Eastern Himalayan ranges being at a somewhat lower latitude than parts of the Western Himalayas, are relatively warmer and the timberline, alpine flora and snowline are at slightly higher altitudes than in the Western Himalayas. More important is the humidity factor, precipitation being heavier in this part of the Himalayas. About 4,000 species of flowering plants are estimated to occur in this region, of which 20 are palms. Palms of this zone are species of *Wallichia, Licuala, Caryota, Daemonorops, Phoenix* and *Pinanga*, Sal. when present, occurs chiefly on the ridges, the intervening depressions being under mixed forests often with an abundance of *Dendrocalamus hamiltonii* Nees. At higher altitudes of this zone there appear two trees, *Betula cylindrostachys* Gamble and *Alns nepalensis* D. Don.

The temperate zone of the Eastern Himalayas extends from 1,524 to 3,657 m. In the lower belt of this zone below 2,742 m. occur a large number of different broad-leaved species, including *Quercus lamellosa* Smith, *Q. lineata* Bl., *Q. pachyphyla* Kurz, and other oaks, *Castanopsis, Michella excelsa* Bl., *Magnolia* and other Magnoliaceae, *Bucklandia populnea* R. Br., *Cedrela*, many laurel's and maples, alder, birch *Pyrus, Symplocos, Echinocarpus, Elaeocarpus, Meliosma* and *Eurya*. Conifers occur mostly above 2,742 m. They are *Abies webbiana* Lindl., *Picea spinulosa* Beiss., *Larix griffthiana* Hort. ex Carr., *Tsuga brunoniana* Carr. and two junipers. Among other plants of this zone may be mentioned numerous rhododendrons and dwarf willows. The bamboos *Arundinaria recemosa* Munro forms a dense growth in places. Two palms also occur in this zone. One is a scandent rattan (*Plectocomia himalayana* Griff.) and the other fan palm (*Trachycarpus martiana* Wendl.)

The alpine zone extends from 3,657 m. to about 4,876 m. Several species of rhododendrons occur here and junipers of the upper temperate zone also extnd high into this zone.

Western Himalayan Region : This comprises the sub-Himalayan tract and the Himalayan range from Kumaun to Kashmir. In general, the Western Himalayas are much cooler and drier than the Eastern. The rainfall varies from 100 to 200 cm., although in some parts of the submontane tracts it reaches 250 cm. or more. These inner valleys and the north-western areas of this region have a dry climate. Naini Tal, Mussoorie, Simla and Kashmir fall under the Western Himalayan region.

The submontane zone and lower hills, up to 1,524 m. contain an almost continuous belt of sal forest in the eastern portion of the tracts as far west as the Yamuna river and to a very small extent beyond. Savannah lands break up the sal belt at intervals. In the western part of this region the forest becomes drier in character, the prevailing species in the submontane tracts and outer hills being *Acacia modesta* Wall., *Olea ferruginea* Royle, *Carissa spinarum* L., *Dodonaea viscosa* L. and other xerophytie species. Among palms, only fiver species occur in contrast to several in the Eastern Himalayas. They are locally found in Kumaun and are *Phoenix sylvestris* Roxb., *P. acaulis* Roxb., *P. humilis* Royle, *Wallichia densiflora* Mart. and *Calamus tenuis* Roxb.

The temperate zone, extending from 1,524 m. to 3,657 m. contains extensive forests of conifers and broad-leaved temperate trees. In the lower elevations, *Pinus roxburghii* Sar. prevails. Soon it gives place to deodar (*Cedrus deodara* Loud.) and blue pine (*Pinus wallichiana* A.B. Jack); higher up, spruce (*Picea morinda* Link.) and silver fir (*Abies pindrow* Spach.) make their appearance and form forests of large extent between 2,438 and 3,352 m. Of other conifers, the yew (*Taxus baccata* L.) is common in some localities. Cypress (*Cupressus torulosa* D. Don) is found locally, and the edible pine (*Pinus gerardiana* Wall.) occurs in the dry inner valleys. Oaks, chiefly *Quercus incana* Roxb., *Q. dilatata* Lindl., and *Q. semecarpifolia* Smith, maples (*Acer* sp.), horsechestnut (*Aesculus indica* Colebr.), poplar (*Populus ciliata* Wall.), elm (mostly *Ulmus wallichiana* Planch), alder (*Alnus nepalensis* D. Don and *A. nitida* Endl., the latter descending below the zone), birch (*Betula alnoides* Ham.), *Cornus, Prunus cornuta* Wall., *Rhododendron arboreum* Sm., and other tres occur. One species of *Trachycarpus* is the only palm occurring in the temperate region and is confined to Kumaun and Garhwal.

The alpine zone extends from the upper limit of the temperate zone to about 4,572 m. or sometimes higher. The charcteristic trees of this zone are the high-level silver fir, the silver birch (*Betula utilis* D. Don), and junipers. Unlike the Eastern HImalayan region, rhododendrons are far less numerous, being represented by only three species.

Andaman Region : The flora of the Andaman and Nicobar Islands is related with that of Burma and Malaysia. The hilly tracts do not exceed 731 m. in height. Among palms, *Daemonorops manii* Becc., and *D. kurzianus* Hook. f. are endemic in the Andaman Islands. Four species of palms are endemic in the Nicobar Islands. They are *Ptychoraphis angusta* Becc., *Bentinckia nicobarica* Becc., *Calamus nicobarics* Becc., and *C. unifarius* Wendl. *Calamus andamanicus* Kurz, and *Pinanga manii* Becc. are endemic in both the Andaman and Nicobar Islands.

The main types of forests in this reigon are mangrove forests, beach forests, evergreen, semi-evergren and deciduous forests and diluvial forests. The mangrove forests are similar to those found on the Indian mainland.

Chapter 27
Some Alien Flowering Plants

With land connections on three sides, north, east and west India has acquired a number of plants of other countries. The Himalayan range in the north has no doubt been a barrier; but there are passes in the north-west of Pakistan and north-east of India and they make a limited spread of plants possible. The areas that have contributed most to the alien elements in the Indian flora are Burma, Malaysia, South-west China, Eastern China, West Asia and Africa.

Hooker (1855) recognized the following principal elements in the Indian flora : (1) the Malaysian element which is the most dominant; (2) the European-Oriental element; (3) the African element; (4) the Tibeto-Siberian element; and (5) the Sino-Japanese element. The subcontinent was considered essentially a meeting place of floras from the west, north and east, and with little botanical character of its own. The then known endemic element was small but we have already shown that this is incorrect Peninsular India and the Himalayas show a high degree of endemicity and hence of floristic distinctiveness. It may be added that almost all the nineteenth century botanists including Hooker obtained a biased view of tropical flora, starting as they did with the improverished temperate floras of their own countries. The flora of Europe was taken as a starting-point with which floras of newly explored tropical lands were compared. Wulff (1950) said : "Thus Hooker (1855) finds in the flora of the Himalayas and Northern India a European element only because the species forming it are likewise found in the flora of Western Europe, although, of course, it is perfectly clear that the centres of areas of these species lie precisely in the Himalayas, whence in post-glacial times they spread to Europe. Hence, this element might be designated as Himalayan in the flora of Europe but in no case as a European element in the flora of India."

Ridley (1942) made certain observations on the Indian flora with reference to the ancient Oligocene flora of the world. The Magnoliaceae, Lauraceae, Hamamelidaceae, Cupuliferae, Salicaceae, Ranunculaceae, Berberidaceae, Hypericaceae, Ternstroemiaceae, Rosaceae, Umbelliferae, Cornaceae, Primulaceae, Styracaceae, Gentianaceae, Boraginaceae, Chenopodiaceae, *Engelhardtia, Carex* and some other plants had a wide distribution in the Cretaceous times over the entire northern part of the world up to the arctic region. In India, this flora is now practically limited to the Himalayas. Rarely, however, some representatives of these families are seen as far southwards as the mountains of Java and Sumatra.

A considerable flora in the tropical parts of India, consisting chiefly of trees and shrubs of the rain forest type, extends from Malaysia through India to North Africa and reappears in Eastern South America. This flora seems to have originated in the Oligocene period, or at least was in existence even then as far north as Sourthern Europe. It has now largely disappeared in the heavily populated and long cultivated regins of India and Central Africa and only some relics are preserved in remote mountainous areas. We have additional proofs of the lost forest flora of the Indo-Ganga plain in the presence of the peacock and the ape, both forest-lovers, and preserved only for religious motives. Among the genera of this category we may mention : *Tetracera, Salmalia, Eriodendron, Sterculia, Buettneria, Erythroxylon, Zizyphus, Casearia, Buddleia, Vitex, Tragia, Elatostemma, Burmannia, Xyris, Sciophila* and many others. *Rhipsalis* is interesting as the only representative of the Cactaceae in the old World.

After the Miocene the eastern end of the Mediterranean Sea became closed up, desert formation occurred in Arabia and Baluchistan (Pakistan) and a desert flora invaded Sind (Pakistan) and Rajasthan and even as far down as the south. Very characteristic of such desert plants are the Salvadoraceae (*Azima* and *Salvadora*), *Dodonaea, Acacia,. Heliotropium, Indigofera,* Crassulaceae, Zygophyllaceae, Capparidaceae and many desert Cruciferae and grasses. Many of the weeds of cultivation common in waste ground all over India are probably due to the invasion from this region.

A number of plants found in Ceylon have come along the South of Asia but have not reached India proper. They show a former land connection between the Malay Islands—probably Sumatra and Ceylon. Of these, the most interesting are *Acrotrema, Anaxagorea, Cullenia, Kurrimia, Pometia pinnata* Forst f., *Timonius koenigii* Bl., and *Lagenophora billardieri* Cass. Most members of the Annonaceae and Menispermaceae of Ceylon have affinites with those of Malaysia and not India.

A large proportion of the maritime plants of India seem to have evolved in the coral islands of Polynesia and Malysia. Their seeds must have drifted along the south coast of Asia, settling on the shores of Ceylon and the Coromandel Coast of India. Some passed on even to East Africa and its islands. Such plants are *Calophyllum inophyllum*L., *Ochrocarpus, Samadera., Xylocarpus granatum* Koen., *Colubrina asiatica* Brongn., *Desmodium umbellatum* DC., *Derris uliginosa* Benth., *Intsia bijuga* O. Ktze., *Rhizophora candelaria* DC., *Bruguiera sexangula* Pers., *Pemphis, Scyphiphora, Guettarda, Wedelia biflora* DC., *Ochrosia, Tournefortia argentea* L. f., *Avicenia officinalis* L., *Cassytha, Hernandia, Flagellaria, Remirea maritima* Aubl. and *Spinifex littoreus* Merr., *Heritiera* and a species of *Dolichandrone, D. spathacea* K. Schum., have travelled from the Indian region as far as the Philippines and New Caledonia by sea.

The genus *Scaevola* is mainly Australian. Two of its species, *S. frutescens* Krause and *S. plumieri* Vahl, have spread to the coasts of Peninsular India.

The palm, *Nipa fruticans* Wurmb., is now found in the Bay of Bengal and along Malaysia as far as the Solomon Islands, but is quite absent from Africa and America. In Eocene time, an almost identical species was abundant in Southern England and along the Mediterranean as far as Cairo. *Acanthus ilicifolius* L. and *A. volubilis* Wall. of the Sundarbans and coasts of India has a closely related species in the Isle of Wight in the Oligocene time.

Ridley (1942) states : "No story of plant distribution is complete without a considerable knowledge of tertiary palaeobotany nor can be understood without a comprehension of the position and form of land surface during that period, the time of the evolution of flowering plants. The modern Asiatic flora is probably what remains of the Oligocene flora which probably occupied all tropical regions as far north as Europe." Many of the early genera and perhaps families have disappeared on account of the vicissitudes of climatic and geological changes, but some species of that date seem to ahve persisted, with little or no alterations, to the present day. Further researches are needed to correlate the extinct and the living groups of flowering plants and probably to fix the date and place of origin of the present Indian flora, and indeed of the floras of the world at large.

Some 38 per cent of the Indian flowering plants have immigrated from foreign lands at various times in the past and have since become naturalized. Some of these aliens have become so well naturalized and successful as to appear really indigenous.

A number of plants originally under cultivation in gardens and fields are now found as escapes that have become thoroughly established by natural agencies like wind, water and animals. Typical examples are : *Lantana camara* var. *aculeata* Mold., *Ipomoea angulata* Lam., *Ageratum conyzoides* L., *Eupatorium glandulosum* H.B. et K., *Helianthus annuus* L., *Tithonia tagetiflora* Desf., *T. diversifolia* A. Gray, *Barleria cristata* L., *Adhatoda vasica* Nes, *Clitoria ternatea* l., *Jatropha gossypifolia* L., *Pedilanthus tithymaloides* Poir., *Eichhornia crassipes* Solms., *Peperomia pellucida* H.B. et K., *Cryptostegia grandiflora* R. Br., *Agave augustifolia* Haw., *Opuntia cochinellifera* Mill. and *Dioscorea alata* L.

Man has been responsible for the import of many alien plants, in some instances deliberately, in others by chance. History tells us that the early Aryan settlers from the countries lying north-west of India brought with them a number of economic plants. It is quite likely that several other plants travelled with them as camp followers, with their seeds or other propagules either mixed with those of the economic plants or stuck to the bodies of the sheep and cattle which these pastoral tribes from the north brought with them. Thus, from almost the dawn of civilization man has been changing the flora of this country byclearing the ground of its primary vegetation and introducing aliens by accident or on purpose.

The great majority of naturalized plant aliens are troublesome weeds competing with cultivated plants or otherwise affecting human welfare. Only in a few instances it is possible to date the actual arrival of a species.

Some weeds have been introduced in comparatively recent times. They have come with food grains, ballast, packing materials and seeds of economic plants, or merely by adherence to the clothing of man and the hair of domestic animals. When they arrived in their new homes, they were further distributed by natural causes like wind and water. Their naturalization might have been further by deforestation, faulty methods of pasturage and harvesting, shifting cultivation, construction of roads and railway lines, and continued sowing of impure seed. *Croton bonplandianum* Baill. is a familiar example of a weed which is now widespread in this country. About the year 1897 a ship arrived at Chittagong (now in East Pakistan) from La Plata with a ballast of mud from South America. To get rid of the mud, it was supplied to a local gardener for soil. That mud contained seeds of this plant; they germinated, and the seedlings flowered and fruited. The plant gradually travelled along railway lines and by steamer to Calcutta and is now a common weed chiefly along rail tracks, roads and canal banks. A similar history attaches to the introduction of *Eupatorium odoratum* L. from the West Indies to India, East Pakistan and Burma, by seeds confined to the ballast heaps of cargo boats calling at Singapore. During recent years it has been seen in the teak plantations of the Kerala State. It is believed that the seeds were brought down to Kerala from Assam by labourers returning from the Assam front after the Second World War. The seeds stuck to their beddings and clothes, thus bridging the long distance from Assam to Kerala. The seeds of *Argemone mexicana* L., now widespread in the tropical parts of the world, are said to have been borne to distant places in ship's ballast from Mexico and the West Indies. Recently, another species of this genus, *A. ochroleuca* Sweet sub-sp. *Ochroleuca*, has also been found in India. *Aeschynomene americana* L., a native of the West Indies and tropical America, has been recorded from the Hazaribagh area. It is possible that a few viable seeds of this species might have reached India along with some packing material during the Second World War, when a large number of American army units were stationed in various parts of India.

Lantana camara var. *aculeata* Mold. and *Eichhornia crassipes* Solms are two remarkable examples of plants which were wilfully introduced to this country as ornamentals but subsequently became so well naturalized as to become serious pests. *Lantana* grows rapidly to form impenetrable thickets in forest lands, and these are difficult to clear by hand because of the prickles on the branches of the plants. Birds feed on the fruits and account for the rapid dispersal of the seeds. *Eichhornia crassipes* Solms (water hyacinth) is a native of the Amazon region of Brazil but is now widespread in Tropics throughout the world. It was brought into India towards the end of the 19th century and soon established itself so successfully that it became a nuisance. Hence its nickname, "Terror of Bengal". It grows gregariously, floating on water or rooted when stranded on wet soil. Rapid vegetative propagation accounts for its profusion, although seeds are also produced. The thick growth of the plant hinders navigation in water channels, chokes drainage, and provides a breeding ground for mosquitoes that spread malaria and other diseases. Eradication of this pest by hand is very laborious. In recent years, certain hormones have been effectively used in some countries for the eradiction of this troublesome pest.

The prodigous spread of some of these recent introductions is viewed with awe. To *Lantana* and *Eichhornia* may be added, among others, *Hyptis suaveolens* Poit. which covers whole hill-sides in Western India and other parts of country, and *Acanthospermum hispidum* DC. which has spread over very large areas in Gujarat and other parts of India. These alien new-comers find conditions very suitable for their growth and propagation; at the same time they are free from their natural enemies (insect, animal or plant), which in their native coutnries keep the plants under control. The alien weeds has been imported into India, but generally its enemies have been left behind.

The introduction, spread and eventual control of the prickly pears (*Opuntia*) in India is a long and eventful story. Although no record exists to show when the first alien *Opunita* reached this country, it must have been well before 1800 A.D. since by then it had become widespread in certain parts of India. It is narrated that sailing boats carried the stem of prickly pears to serve as vegetables at a time when anything green, not actually poisonous even if unpalatable, was used to prevent scurvy among sailors, *Opuntia* may thus have found its way from the new World into India via Europe. It is on record that small enclosures, bounded by hedges of *Euphorbia and Opuntia,* caused the entanglement of Tipu's horse in the battle of Poongar on the banks of the river Cauvery on September 12, 1790. The practice of making fences out of prickly pears, and the natural dispersal of their seeds by birds after they had eaten the fruits, greatly contributed to the spread of the plants. A long fence of this kinds, called the "Salt Wall", was made over miles of the Rajasthan border to prevent smuggling. Prickly pears were also used for the protection of young shade trees along roadsides. All this afforded the plants many new starting points for fresh encroachment and they gradually became a serious pest of garden and field. In some places they formed dense thickets, an excellent shelter for snakes. At the end of the 18th century, the East India Company tried to establish in this country a cochineal dye industry which was till then a Spanish monopoly in America. The cochineal insect feeds on species of *Opuntia* and through the efforts of the Company, in colaboration with Dr. James Anderson at Madras and Dr. William Roxburgh at Calcutta, several new species of *Opuntia* were introduced into cultivation in India, along with the cochineal insect. However, the production of the dye was not satisfactory and by 1810 the Government was compelled to discontinued till 1883. The insect, introduced into India in 1795, was a blessing in disguise. It spread rapidly on plants of *O. monacantha* Haw. devouring them branch and root, and in the course of twenty years annihilated this pest in South India almost totally. In the north, it took about sixty years for the insect to spread from Bengal up the Ganga plain and over the Indus plains more than 1,200 km. away. *Opuntia monacantha* Haw. had by then become a pest in the Punjab. Sher Singh, ruler of Lahore, inflicted fines on people who allow this plant to grow on their grounds. Within a year of the invasion of the Punjab plain by the cochineal insect, the prickly pear in the area was thoroughly destroyed and a large supply of dye was available to the local Kashmiri dyers. *Opuntia monacantha* Haw. is now comparatively rare. *Opuntia dillenii* Haw. and *O. elatior* Mill., on the other hand, were immune to the species of coachineal insect originally introduced into India. In comparatively recent times, other species of the cochineal insect were imported into South India in an attempt to eradicate *Opuntia dillenii* Haw. in particular. This measure was successful in bringing this pest also under control.

Bibliography

Abdi, R.D., 1993. Naldharis of Saurashtra; A Glimpse into Their Past and Present, Suchitra Offset, Bhavnagar, Gujarat.

Alfred, J.R.B. 1990. Taxonomy in Conservation and Management of the Environment, Taxonomy in Environment and Biology (ed. Director, ZSI), 323-330, ZSI, Calcutta.

Alfred, J.R.B. Darlong, V.T. & Sati, J.P., 1989. Wildlife Conservation in North-east India : Problems and Realities. Exposure (2(1): 11-17.

Alfred, J.R.B., Darlong, V.T., Hattar, S.J.S. & Paul, D., 1989. Mocro-arthropods and their conservation in some North-east India soil, In : Advances in Management and Conservation of Soil Fauna (Ed. G.K. Veeresh, D. Rajagopal and C.A. Viraktamath), 309-320. Oxford & IBH, New Delhi.

Ali, S. & Ripley, S.D. 1983. A Pictorial Guide to the Birds of the Indian Subcontinent. Bombay Natural History Society & Oxford University Press, India, p. 187.

Anon 1979. Tropical Grazing Land Ecosystems, A state-of-knowledge Report UNESCO. UNEP/FAO.

Anon. 1992. Jhum : Is There A Way Out? The Price of Forests (Ed. A. Agarwal), 304-310, CSE, New Delhi.

Anonymous, 1979. Tropical Grazing Land Ecosystems. A state of knowledge report prepared by UNESCO/UNEP/FAO.

APHA, 1985. American Public Health Association. Standard Method for the Examination of Water and Waste Water. APHA Inc, New York.

Arora, G.S. and Julka, J.M. (1993). Status report on biodiversity conservation: Western Himalayan ecosystem. IIPA, New Delhi (in press).

Askins, R.A. (1995). Hostile landscapes and the decline of migratory songbirds: Science, 267: 1956-1957.

Awasthi, D.D., 1960. Contributions to the Lichen Flora of India & Nepal. JIBS. Vol. XXXIX.

Bell, R.A.V., 1971. A Grazing Ecosystem in Serengeti. Scientific American, 225, 86-93.

Berwick, S.H. 1974. The Community of Wild Ruminants in the Gir Forest Ecosystem, India. A Thesis for the Degree of Doctor of Philosophy, p. 226.

Beddome, R.H. 1863. The Ferms of the Southern India. Madras Presidency.

Bhatnagar, G.P. 1984. Limnology of Lower Lake Bhopal with Special Reference to Sewage Pollution. MAB Project (Report) pp. 1-74.

Bhatnagar, G.P. 1992. Studies of Upper Lake and Water Works Ecosystems of Bhopal, MAB Tech. Report. pp. 1-81.

Bhatnagar, G.P. 1995. Eutrophication in India Tropical Lakes with Special Reference to Bhopal. Shrivasta, O.N. (Ed.) Recent Advances in Limnology in India, Ch. 61. Today and Tomorrow Publishers, New Delhi (In press).

Biswas, S. & Ghosh, A.K., 1976. Impact of Shifting-cultivation on Wildlife in Meghalaya. In : Shifting Cultivation in North-east India (Ed. B. Pakem *et. al.*), 77-79, NEISSR, Shilong.

Blackman, F.F., 1956 Influence of Light and Temperature of Leaf Growth. In: Mithore., F.L. (ed.). The Growth of Leaves. Butter Worths, London, 159-169.

Blatter, E & d' Almeid, J.F. 1922. The Ferms fo Bombay, Bombay.

Bloom, D.E., 1995. International Public Opinion in the Environment Science, 269, 354-357.

Boergesen, F., 1930 Some Indian Green and Brown Algae, JIBS. Vol. IX.

Brain Groombridge. 1992. Global biodiversity: Status of Earth's Living Resources. World Conservation Monitoring Centre, Cambridge.

Bor, N.L., 1953. Manual of India Forest Botany. Bombay.

Bruhl, P.A., 1931. A Census of Indian Mosses. Records of the Botanical Survey of India. Vol. XIII. No. 1.

Bubble, S.P. & R.B. Foster, 1986. Commonness and Rarity in Neotropical Forests: Implications for Tropical Tree

Conservation. Pages 205-231 in M.E. Soule editors Conservation Biology: The Science of Scarcity and Diversity, Sinauer Associates, Inc. Publ. Sunderland, Massachusetts.

Butler, E.J., 1905. Some Indian Forest Fungi. The Indian Forester. Vol. XXXI. No. II.

Burley, J. 1994. World Forestry: The Professional Scientific Challenges. The Leslic L. Schaffer Lecxtureship in Forest Science, Vancover, B.C. Canada.

Bruhl, P.A., 1931. A Census of Indian Mosses. BSI. Vol. XII. No. I.

Castri, F. Di, J.R. Vernhes, & T. Younes, 1992 (Eds.) Inventorying and Monitoring Biodiversity. Pages 1-27 in Report of an ad hoc group of IUBSSCOPE-UNESCO, Programme Function of Biodiversity (30-31 Jan.), Paris.

Calder, C.C. 1937. An outline of the vegetation of India. ISCA. Calcutta.

Chaterjee, D., 1939. Studies on the Endemic Flora of India and Burma. The Journal of the Asiatic Society of Bengal. Vol. V. : 19-64.

Champion, H.G. & Seth, S.K., 1986. A Revised Survey of Forest Types of India, FRI, Dehradun.

Chaterjee, S., 1995. Global 'Hot Spots' of Biodiversity. Current Science, 68, 12, 1178-1179.

Chavan, S.A., Gogate, N.S. & Patel. C.D. 1991. Endangered Ecosystem of schoolpanesjwar Sanctuary, Rajpipla, Gujarat, Paper presented at Session of Indian Science Congress, Vadodra, India.

Choudhury, N. (1991). Status of wild elephants (Elaphas maximus Linn.) in Cachar and Northe Cachar hills, Assam–a preliminary investigation. J. Bombay Nat. Hist. Soc. 88(2): 215-221.

Chopra, R.S. 1932. Liver worfts of the Western Himalayas and the Punjab Plains. Pf. II Lahore.

Chouhan, A.S. & Singh, D.K. 1989. Changing patterns in the flora due to deforestation. Environmental conservation and Wasteland Development in Meghalaya (Ed. A. Gupta & D.C. Dhar), 75-107. Meghalaya Science Society (MMS), Shillong.

Cody, M.L., 1986. Diversity, Rarity, and Conservation in

Mediterranean Climate Region. Pages 112-152. In M.E. Soule editors Conservation Biology of Science of Scarity and Diversity. Sinanuer Associates, Inc. Publ. Sunderlanf Massachusetts.

Clarke, C.B. 1880. A Review of the Ferns of Northern India. London.

Cohen, J.E., 1995. Population Growth and Earth's Human Carrying Capacity Science, 269, 341-346.

Craig, J.R., Vaugha, D.J. and Sikinner, NB.J., 1988. Resources of the Earth. Prentice-Hall, Eagle wood Cliffs, N.J.

Daily, G.C., 1995. Restoring Value to the World's Degraded Land. Science, 269, 350-354.

Daily, G.C., Ehrilich, P.R. and Haddad, N., 1993. Double Jeystone Bird in A Keystone Species Complex. Proc. Natl. Acad. Sci., USA, 90, 592-594.

Daniel, J.C. 1983. The Book of Indian Reptiles. Bombay Natural History Society and Oxford University Press, India 141 p.

Daniels, R.J., Ranjit, M. Hegde, N.V. Joshi, M. Gadgil, 1991. Assigning Conservation Value: A Case Study from India. Conservation Biology 5 (4): 464-475.

Daniels, R.J.R. Hegde, M. and Gadgil, M. 1990. Birds of the man made cosystem: The plantations. Proc. Indian Acd. Sci. (Anim. Sci.) 19(1) 79-89.

Darlong, V. T., Choudhury, D., Sati, J.I. & Alfred, J.R.B. 1989. Wildlife and the Vanishing Forests: An appraisal. In: Environment Conservation and Wasteland Development in Meghalaya (Ed. A Gupta and D.C. Dhar), 108-127, Meghalaya Science Society, Shillong.

Darlong, V.T. & Alfred, J.R.B. 1982. Difference in Arthopod Population Structure in Soils of Forest and Jhum Sites of North-east India. Pedobiologia 23: 112-119.

Darlong, V.T. & Alfred, J.R.B. 1993. Micro-arthropod diversity in some soils of North-east India with special reference to effect of shifting cultivation. In: Himalayan Biodiversity: Conservation Strategies (Ed. U. Dhar), 331-323. GBPIHED, Almora.

Darlong, V.T. & Alfred, J.R.B., 1989. Effect of Shifting Cultivation (Jhum) on Soil Fauna with particular reference to the

Earthworms in North-East India. In advances in Management and Conservation of Soil Fauna (Ed. G.K. Veersh, D. Rajagopal & C.A. Viraktamath), 299-308, Oxford & IBH, New Delhi.

Darlong, V.T., Choudhury, D., Sati, J.P. & Alfred, J.R.B. 1989. Wildlife and the Vanishing Forests: An Appraisal. Environmental Conservation and Wasteland Development in Meghalaya. (Ed. A. Gupta & D.C. Dhar), 108-126, MSS, Shillong.

Deb Roy, S. & Jackson, P. 1993. Mayhem in Manas: the Threats of India's Wildlife Reserves. The Law of the Mother (Ed. E. Kermf), 156-161, Sierra Club Books, San Francisco.

Desikachary, T.V. Cyanophyla. 1959. The Indian Council of Agricultural Research, New Delhi.

Ehrlich, P.R. 1994. Energy use and biodiversity loss. Phil. Trans. R. Soc. Lond. B 334, 99-104.

Elmes, G.W. and Thomas, J.A. 1992. Complexity of species conservation in managed habitats: Interaction between Maculinea butterflies and their ant hosts. Biodiversity and Conservation 1: 155-169.

Erhardt, A. 1985. Diurnal lepidoptera: Sensitive indicators of cultivated and abandoned grasslanf. J. Appl. Ecol. 22: 849-861.

Flint, Michael, 1991. Biological Diversity & Developing Countries. A Synthesis Paper. ODA.

Gadgil, M. and Meher-Homji, V.M. 1990. Ecological diversity. In: J.C. Daniel and J.S. Serro ed. Conservation in developing countries: Problems and Prospects. Proceedings of the Centenary Seminar of the Bombay Natural History Society: pp. 175-198.

Gadgil, M., 1989. Forest Exploitation : Patterns and Pricesses. Biodiversity. Biovigyanam. 15, 1, 1-21.

Gadgil, M., 1993. Forestry with A Social Purpose. In : People's Rights and Environmental Needs. W. Fermandes and S. Kulkarni (Eds.) Indian Social Institute, New Delhi, pp. 111-130.

Gadgil, Madhav, 1994. Reckoning with Life. The Hindu Survey of the Environment.

Ghosh, A.K. & Tiwari, K.K. 1984. Faunal Résources of North-east India. In: Resource Potentials of North-east India Vol. II (Living

Resources Ed. R.S. Tripathi), 105-109, Meghalaya Science Society, Shillong.

Ghosh, A.K. 1991. Animal Resource of India: Protozoa to Mammalia, State of the Art Report. Zoological Survey of India. Ministry of Environment & Forests, Vol. XI, XXVII.

GOI, 1984. Conservation of Biological Diversity in India: An Appraisal. Ministry of Environment and Forests, p. 13.

GOI, 1992. Tradition, Concerns & Efforts in India. National report to UNCED, Ministry of Environment & Forests, Govt. of India.

Golterman, H.L. & Clymo, R.S., 1978. Methods for Physical and Chemical Analysis of Fresh Water. Blackwell Scientific Publications, Oxford, 213 pp.

Government of India. 1994. Conservation of Biological Diversity in India: An Approach. Ministry of Environment & Forests, Govt. of India, New Delhi, pp. 48.

Govt. of India. 1993. The State of Forest report - 1993. Forest Survey of India, Dehradun.

Govt. of India. 1994. Conservation of Biological Diversity in India: An Approach. Ministry of Environment and Forests (MOEF), New Delhi.

Green, J.B., 1993. Natural Resources of the Himalaya and the Mountains of Central Asia. Pages 137-290, IUCN Publ.

Gupta, A. 1989. Impact of Deforestation on the Aquatic Fauna of Meghalaya. Environmental Conservation and Wasteland Development in Meghalaya (Ed. A. Gupta & D.C. Dhar), 128-139, MSS, Shillong.

Gupta, P.N. & S. Cgandra, 1987-89 to 1978-88. Forest Working Plan Pithoragarh Forest Division (Part 1 & 2).

Harley, J.L. & Smith, S.E. 1983. Mycrorhizal Symbiosis. Academic Press London.

Harrison, J., Miller, K., McNeely, J., 1982. The World Converage of 'Protected Areas. IUCN World National Parks Congress, Bali-Indonesia.

Henning, D.H. and Mangum, W.R. 1989. Managing the Environmental Crisis. Durham University Press, Durham & London.

HMG, 1988. Master Plan for the Forestry Sector 1988. His Majesty's Government of Nepal.

HMG, 1991. Statistical Hand Book of Nepal 1991. Central Bureau of Statistics, HMG, Nepal.

HMG, 1993. Statistical Hand Book of Nepal 1993. Central Bureau of Statistics, HMG, Nepal.

Hooker, J.D., 1904. A Sketch of the Flora of British India. London.

Ishwaran, N. 1992. Biodiversity, Proected Areas & Sustainable Development. Nature & Resources, Vol, 28 No 1, 1992.

IUCN/UNEP/WWF, 1991. Caring for the Earth. A Strategy for Sustainable Living, Glamd, Switzerland.

IUCN *et. al.* 1990 1988. Conserving the World's Biodiversity. ICUN/ WRICI/WWFUS/World Bank.

IUCN, UNEP and WWF, 1980: World Conservation Strategy.

Jain, S.K., 1994. Biodiversity: Some Perspective in Study and Conservation Reg. Conv. Min. Env. For., G.I. Lucknow.

Java, R.L. 1990. Environment and Wildlife Conservation in Gujarat State. Status Paper Forest Department, Gujarat State.

Java, R.L. 1990. Environment and Wildlife Conservation in Gujarat State. Status Paper. Forest Department, Gujarat State.

Jonshingh, A.J.T. 1986. Diversity and conservation of carnivorous mammals in India. Proc. Indian Acad. Sci. (Anim Sci/Plant Sci.) Suppl.: 73-90.

Jorgensen, S.E., 1990. Erosion and Filtration. In: Gudelines of Lake Management (Chap. 3) pp. 13-19. ILEC, Japan.

Kashyap, R.S. 1929. Liverworts of Western Himalayas and the Punjab Plains, Pt. 1. Lohare.

Kaul, V., 1971. Production and Ecology of Some Macriphytes of Kashmir Lakes. Hydrobiologia, 1(2): 63-69.

Kernan, H., Bender, W.L. and Bhatt, B.R. 1986. Report of the forestry private sector study USAID Mission to Nepal. Kathmandu, Nepal.

Khanchandani, M.S. & Sinha S.K. 1973. Working Plan for the Forests of Panchmahals and Kaira (Balasinor) Districts. Forest Department, Gujarat State.

Khanna, P; Srivastava, A.K. 1995. Gatt-Before it is too Late. Indian Forester, March, 1995.

Khoshoo, T.N., 1986. Environmental Priorities in India and Sustainable Development. Presidential Address, 73rd Session, India Science Congress Association, Jan. 3-7, 1986, New Delhi.

Khoshoo, T.N., 1992. 2nd Pdt. Govind Ballabh Pant Memorial Lecture. G.B.P.I.H.E.D., Kosi Almora, India.

Khoshoo, T.N. 1994. Census of India's Biodiversity: Taks Ahed. Curr. Sci. 67; 557-582.

Krishna Kumar, Asha, 1994. Hamessing A Heritage. Forntline March 11, 1994.

Lal,B., Kardam, S.C.L. and Jatav S.R., 1993. Depressed class Research, IARI, New Delhi.

Lal, B., 1991. Tribal India/shifting cultivation: Amelioration of problem soils and rea. IGFRI Jhansi, India.

Lal, B. & Chavan, S.A. 1994. Grasslands of Saurashtra and Kachchh: Status and Future Strategy for Management of Lesser Florican (*Sypheotides Indica*) Habitat. Paper presented at Lesser Florican Conservation Workshop held at Vadodara. 1994.

Lal, B. & Chavan, S.A. 1995. Moisture Conservation Strategy in Drought Prone Areas: Technological Alternatives to Enlist People's Participation. Wasterlands News, XI (1) 23-27.

Lal, B. & Chavan, S.A. 1994. Grassland of Saurashtra and Kachchh; Status and Future Strategy for Management of Lesser Florican (Sypheotides indica) Habitat. Paper presented at lesser florican conservation workship held at Vadodra, 1994.

Lal, B. 1995. Greater Gir Ecosystem, Recapturing of Lost Territory by *Panthera leo persica* [Asitatic Lion] in India.

Lal, B. 1995. Draft Working Plan of Jamnagar Forest Division. Forest Department, Gujarat State.

Li, C.Y., Massicote, H.B. & Moore, L.V.H., 1992. Nitrogen Fixing Bacullus Sp Assoicated with Douglas - Fire Tuberculate Ectomycorrhizae. Plant and Soil 104: 35-40.

Lowe-Mc Conell, R.H. 1987. Ecological studies in tropical fish communities Cambridge University Press.

Mac Coy, D.E. & H.R. Mushinsky, 1992. Rarity of Organisms in the

Sand Pine Scurb Habitat of Floira. Conservation Biology 6(4): 537-548.

Mann, C.C. 1995. Filling in Florida's gaps: Species protection done right? Science 369: 318-320.

Maheshwari, J.K. 1962. Studies on the naturalized flora of India. Proceeding of the Summer School of Botany, Darjeeling. (1960). New Delhi.

Mathur, C.M. 1960. Forest Types of Rajasthan. The Indian Forester, Vol. 86. No. 12.

Marak, T.T.C. 1989. Causes of Deforestation with special reference to Meghalaya. Environmental Conservation and Wasterland Development in Meghalaya (Ed. A. Gupta & D.C. Dhar), 53-58, MSS, Shillong.

May, R.M., 1992. How Many Species Inhabit the Earth? Scientific American, Oct., 18-24.

McNeely, J.A. 1988. Economics and Biological diversity, IUCN, Switzerland.

McNeely, J.A. 1994. Lessons from the Past: Forests and Biodiversity. Biodiversity and Conservation 3: 3-20.

Medley, K.E., 1993. Extractive Forest Resources of the Tana River National Primater Reserve, Kenya. Economic Botany 17(2): 171-183.

Meyers, N. and Ayensu, E.S., 1983. Reduction of Biological Diversity and Species Loss. Ambio, 12: 72-74. Harshberger, J.W., 1895. Some new ideas. Philadelphia Evening Telegraph.

Mishra, R. 1968. Ecology Work Book. Oxford, New Delhi, IBH Publishing Company.

Mishra, R., 1968. Ecology Work Book. Oxford, New Delhi, IBH Publishing Company.

Mitten, W. 1861. Music Indane Orientalies (Supp. Bot. Vol. 1) J.P. LS. Vol. V.

Mlot, C. 1995. In Hawaii, taking inventory of a biological hot spot. Science 269: 322-323.

Molina, R. and Trappe, J.M. 1982. Applied Aspects of Ectomycorrhizae. In Advances in Agricultural Microbiology (N.S. Subba Rao ed.) pp. 305-324. Oxford and IBH Publ. Co., New Delhi.

Mongia, S.G. & Navroji, R.K. 1984. Birds of Rajpipla Forest-south Gujarat, with Notes on Nests Found and Breeding Recrods and Some New Observations, Journal of Bombay Natural History Society, India 80(3).

Myers, N. 1988. Threatened Biotas: Hot Spots in Tropical Forests. The Environmentalist 8(3): 187-208.

Myers, N., 1988. Tropical Forests and their Species – Going, Going .....? In: Biodiversity, National Academy Press, Washington D.C. pp. 28-32.

Nair, N.C. and Daniel, P. 1986. Floristic diversity of Western Ghats and its conservation. Proc. Indian Acad. (Anim. Sci/Plant Sci) Suppl.: 127-164.

Natarajan, V., 1994. Report on Grassland Ecology Project on Velavadar National Park.

Odum, E.P., 1971. Fundamentals of Ecology. Philadephia, London, Saunders, 3rd ed.

Odum, S.P. 1971. Fundamentals of Ecology. W.B. Saunders, Philadelphia, London, 3rd ed.

Oosting, H.J. 1956. The Study of Plant Communities. Freeman Publishing, Co., San Francisco.

Oosting, H.J. 1956. The Study of Plant Communities. San Francisco, Freeman.

Pickett, S.T.A. and Cadenasso, M.L. 1995. Landscape Ecology: Spatial heterogeneity in Ecological Systems. Science 369: 331-334.

Pimentel, D., Harvey, C., Resousdarmo, P., Sinclair, K., Karz, D., Mcnair, M., Crist, S., Shapritz, L., Fitton, L. and Blair, S.R., 1995. Environmental and Economic Costs of Soil Erosion and Conservation Benefits. Science, 267, 1117-1122.

Pimm, S.L., Russell, G.J., Gittleman, J.L. and Brooks, T.M. 1995. The future of biodiversity, Science, 269: 347-350.

Prakash., I. 1986. Faunal diversity in the arid tracts. Proc. Indian Acad. Sci. (Anim. Sci/Plant Sci.) Suppl.: 45-58.

Prater, S.H., 1980. The Book of Indian Animals. Bombay Natural History Society & Oxford University Press, India 323 p.

Prime, S.L., Russell, G.J., Gittleman, J.L. and Brooks, T.M. 1995. The Future of Biodiversity. Science, 269, 347-350.

Raizada, M.B. & Sahni, K.. 1960. Living Indian Gymnosperms. Pt. I. Vol. V. No. 2.

Rabinowtiz, D., 1981. Seven forms of Rarity. Pages 205-217. In H. Synge Editors. Biological Aspects of Rare Plant Conservation. Wiley, Chechester, U.K.

Ramkrishnan, K. & Subramamian, C.V. 1952. The Fungi of India. J. of Madras. Vol. XXII(B). No. 1.

Ramakrishnana, P.S. 1985. Tribal Man in the Humid Tropics of the North-east Man in India 65(1): 1-32.

Randhawa, M.S. 1959. Zygnemaceae. ICAR. New Delhi.

Raninowitz, D., S. Cairns & T. Dhillon, 1986. Seven Forms of Rarity and Their Frequency in the Flora of the British Isles. Pages 182-204 in M.E. Soule Editors Conservation Biology: The Science of Scarcity and Diversity, Sinauer Associates, Sunderlands, Massachusetts.

Rao, R.R. and Hajra, P.K. 1986. Floristic diversity of the Eastern Himalayas in a conservation perspective. Proc. Indian Acad. Sci. (Anim. Sci/Plant Sci.) Supplement: 103-126.

Rawat, G.S. 1983. Studies on High Altitude Flowering Plants of Kumaun Himalaya. Ph.D. Thesis, Botany Department Kumaun University, Nainital.

Rawat, G.S., & Y.P.S. Pangtey, 1987. A Contributions to the Ethnobotany of Alpine Regions of Kumaun. Journal of Economic & Taxonomic Botany 11: 139-148.

Reid, W.V. and Miller, K.P., 1989. Keepling Options Alive: The Scientific Bans for Conserving Biodiversity. World Resource Institute, Washington, DC.

Ridley, H.N. 1942. Distribution Area of Indian Floras. RBG. Calcutta.

Robinson, S.K., Thompson, F.R. III Donovan, T.M., Whitehead, D.R. Faaborg, J. 1995. Regional Forest fragmentation and the nesting success of migratory birds. Science 267: 1987-1989.

Robinson, S.K., Thompson, F.R., Donovan, T.M., Whitehead, D.R. and Faaborg, J., 1995. Regional Forest Fragmentation and the Nesting Success of Migratory Birds. Science, 267, 1987-1990.

Rodgers, W.A. & H.S. Panwar, 1988. Planning A Wildlife Protected Area Network in India. Vol. I & II, Wildlife Institute of India, Dehradun.

Rodgers, W.A. & H.S. Panwar, 1988. Planning A Wildlife Protected Are Network in India. WII, Dehra Dun, India.

Rodgers, W.A. & Panwar, H.S. 1988. Planning of Wildlife Protected Area Network in India: A Report Wildlife, Institute of India, Dehradun, India 50p.

Rodgres, W.A. & Sawarkar, V.B. 1988. Vegetation Management in Wildlife protected Areas in India. Aspect of Applied Biology. 16, Wildlife Institue of India, Dehradun, pp. 407-421.

Ruckelshaus, W.D. 1994. Towards a Sustainable World. Scientific American, 166-176, September, 1994.

Ruckeishaus, W.D., 1994. Towards A Sustainable World. Scientific American, September, 166-175.

Razi, B.A. 1959. Second List of species and genera of Indian Phanerogams not in J.D. Hooke's Flora of British India. R.B.S.I. Vol. XVIII. No. 1.

Sachs, I. 1992. Transition Strategies for the 21st Century. Nature & Resources, Vol. 28, No. I.

Sahni, KC. 1981. Botanical panorama of Eastern Himalayas. Aspect of change. (ed.) J.S. Lall. New Delhi: Oxford University Press: 32-49.

Samant, S.S., Rawal, & U. Dhar, 1993. Botanical Hot Spots of Kumaun Conservation Perspectives for Himalaya. Pages 377-400. In U. Dhar (Ed.) Himalayan Biodiversity: Conservation Strategies. Gyanodaya Prakashan, Nainital.

Santapau, H. 1958. Systematic Botany of Angiosperms. HBR.

Sankaran, R. & Rahmani, A.R. 1986. Study of ecology certain endangered speceis of wildlife and their habitats. The lesser florican Annual Report 2, BNHS, Bombay.

Sastry, C.A. Aboo, K.M. Bhatia, H.L., & Rao, A.V., 1970. Pollution of Upper Lake and its Effect on Bhopal Water Supply. Env. Hlth 12(3): 218-238.

Sen, S. 1989. Urabanisation and Erosion in Traditional Values and Practices in the tribal Setting of North-east India. Envrionmental Conservation and Wasteland Development in Meghalaya (Ed. A. Gupta & D.C. Dhar), 179-188. MSS, Shillong.

Shah, G.L. 1978. Flora of Gujarat State, Vol. I & II. University Press, Vallabha Vidyanagar, Gujarat, India, 1074p.

Shah, N.C. & S.K. Jain, 1988. Ethno-Medico Botany of Kumaun Himalaya, India, Social Pharmacology 2(4): 359-380.

Sharma, M.M., 1995. Study on Driage of Grass During Departmental Collection. Forest Department, Gujarat State (Unpublished).

Sharma, T.C. 1976. The Prehistoric Background of Shifting Cultivation. Shifting Cultivation in North East India (ed. B. Pakem, J.B. Bhattacharjee, B.B. Dutta & B. Datta Ray), 1-4 NEICSSR, Shillong.

Singh, G.H., Narain P, Babu, R. and Aberol I.P., 1991. Apprasial of erosion problems in India. Anon. Report, CSWCR & TI Dehradun.

Singh, H.S. & Kamboj, R.D., 1995. Draft Biodiversity conservation Plan for Gir, Forest Department, Gujarat State 1995.

Singh, H.S. & Rana, V.J., 1995. Management Plan of Blackbuck National Park Velavadar. Forest Department, Gujarat State

Singh, H.S. 1994. Status Report on Mangrove in Gujarat State. Forest Department, Gujarat State.

Singh, N.P., 1994. Conservation of Biodiversity. Second training course on Management of Natural Resources and Environment, Botanical Survey of India, Ministry of Environment & Forest, Govt. of India.

Sinha, S.K. & Joshi, R.R., 1972. Working Plan for Kutch Division. Forest Department Gujarat State.

Sinha, S.K. & Pinto, A.G., 1972. Working Plan for Junagadh Forest Division. Forest Department, Gujarat State.

Sinha, S.K. 1989. Colourful Wildlife of Gujarat. The Gujarat Experience Forest Department, Gujarat State.

Snader, K. 1990. Overview of Supply Issues (Abs.) in a workshop on "Taxol and Taxus: Current and Future Perpective", NCI, Bathesada, MD.

Stebbing, E.P. 1922-62. The Forests of India. Vol. I-IV London.

Subramanyam, K. 1962. Aquatic Angisperms. CS.IR, New Delhi.

South Asian Association for Regional Cooperation, 1992. Regional Study on the Course and Consequences of Natural Disasters and the Protection and Preservation of the Environment. SAARC, Kathmandu, p. 212.

Sukumar, R., (1989. The Asian Elephant-Ecology and management. Cambridge Unisersity Press. Cambridge U.K. Edition II, 1992.

Swales, S. and West, D.W. 1991. Distribution, abundance and conservation status os native fish in sime waikato streams in North Island of New Zealand. JI. R. Soc. N.Z. 21: 281-296.

Tamot, P. & Bhatnagar, G.P. 1989. Studies in raw Water Quality of Upper Lake and its Change During Various Stages of Treatment At 5 M.G.D. Water Treatment Plant (P.H.E.D.), Bhopal (M.P.). J. Hydrobiol. 5(1): 35-38.

Tamot, P. & Bhatnagar, G.P., 1988. Limnological Studies on Upper Lake Bhopal, Proc. of Nat. Symp. on Present Past and Future of Bhopal Lakes, pp. 37-40.

Tamot, P., 1991. Limnological Studies on the Upper Lake at Intake Point and 5 M.G.D. Water Treatment Plants (P.H.E.D.), Bhopal, Ph.D. Thesis B. University, Bhopal, pp. 1-143.

Thottathri, K., 1962. Contributions to the Flora of Andamans & Nicobar Islands. BSI. Vol. IV.

Terborgh, John, 1992. Diversity and the Tropical Rain Forests. Scientific American Library, NY.

Thursell, J. & J. Harrison, 1992. National Parks and Nature Reserve in Mountain Environment and Development. Geo Journal 27(1): 113-126.

Tomlinson, D. 1992. Look what the stork brought. New Scientist. 135: 44-45.

Trappe, J.M., 1977. Selection of Fungi for Ectomycorrhizal Inoculation in Nurseries. Ann. Rev. Phytopathol. 15: 203-222.

Triathi, R.S. Tiwri, B.K. & Barik, S.K. 1995. Sacred Groves of Meghalaya: Status and Strategy for Their Conservation. NAEB, NEHU, Shillong, p. 125.

Troup, R.S., 1971. The Silviculture of Indian Trees, Vols. I-III. Oxford.

Trivedi, R., Valecha, V. & Bhatnagar, G.P. 1988. Non Homogenous Horizontal Distribution of Faecal Bacterial Indicating Degree of Pollution. Indian J. Applied and Pure Biol. 3: 31.

Trivedi, R., Valecha, V. & Bhatnagar, G.P., 1989. Seasonal variation and Differentiation of Cliform Bacteria in Lower Lake of Bhopal, India Environment and Ecology 7(1): 206-210.

Usher, M.B. 1986. Wildlife Conservation Evaluation: Attributes, Criteria and values. Pagers 3-44. In M.B. Usher, Editors Wildlife conservation evaluation. Chapman and Hall, London.

Valecha, R., Valecha, V. and Bhatnagar, G.P., 1988. Distribution of Some Bacteria (Total Coliform, E. coli and Faecal Streptococci) and their Ratios in the Lower Lake of Bhopal, India J. Applied and Pure Biol. 3(2): 107-113.

Valecha, V., Calecha, R. & Bhatnagar, G.P., 1989. Diurnal Variation in Some Physico-chemical Parameters During Summer in Lower Lake of Bhopal Geobios, 15: 170-173.

Valecha, V., Trivedi, R. & Bhatnagar, G.P. 1987. Biological Assessment of Tropical Status of Lower Lake Bhopal. Poll. Res. 6(3&4): 91-93.

Valecha, V., Valecha, R. & Bhatnagar, G.P., 1991. Limology of Sewage Polluted Lake: The Lower Lake Bhopal, India. Int. Rev. Geo. Hydrobiol. 76(1): 137-47.

Vanclay, J.K. 1992. Species Richness & Productive Forest Management Proceedings of the Oxford Conference on Tropical Forests.

Vollenweider, R.A., 1969. A Manual on Methods for Measuring Primary Production in Aquatic Environment. IMP Handbook No. 12, 213 p.

Wafer, M.U.M. 1986. Corals and coral reefs of India. Proc. Indian Acad. Sci. (Anim. Sci./Plant Sci.) Suppl.: 19-44.

Wallace, A.R. 1970. The Geographic Distribution of Animals. Macmillan, London, 2 Vols.

Waller, W., 1981. Fresh water Marshes. Ecology and Wildlife Management, Univ. Minnesota Press. Minneapolis, MU 146 p.

Waltors, C. 1986. Adaptive Management of Renewable Resources. Macmillan, New York, USA.

Watson, Col. J.W., 1984. Statistical Account of Junagadh, Being Junagadh Contribution to the Kathiawar Portion of the Bombay Gazetter Printed at Bombay Gazettee, Steam Press, Rampart Row, Bombay, p. 160, 1884.

WEC 1986. Executive summary of energy sector synopsis report 1985-86. WEC Report.

Wetzel, R.G. & Likens, G.E., 1979. Limnology Analysis (2nd ed.). Springer Verlag Publ., New York.

Wharton, C.H. 1968. Man, Fire and Wild Cattel in South-east Asia. Proceeding of the Annual Tall Timbers Fore Ecology Conference. 8: 107-167.

Williams, N. 1995. Slow start for Europe's habitat protection plan. Science 269: 220-322.

Wilson, E.O., 1988. The Current State of Biological Diversity. In: Biodiversity. National Academy Press, Washington, D.C., pp. 3-5.

World Bank, 1989. World Development Report. World Bank Publication.

World Conservation Monitoring Centre, 1992. Global Biodiversity: Status of Earth's Living Resources. Champman and Hall, London.

World Conservation Monitoring Centre, 1993. Global Biodiversity. Champman and Hall, London.

World Resource Institute 1989. World Resources 1989-90. Oxford University Press. New York.

World Resource Institute 1990. World Resources 1990-91. Oxford University Press, New York.

World Resource Institute 1992. World Resources 1989-93. Oxford University Press, New York.

World Resource Institute, 1992. World Resources, 1992-93. Oxford University Press, New York.

Zahlbruchner, A. Catalogu Lichenum Universalis, B and 1-10. Leipzig/ Berlin. 1922-1940.